Agricultural Plant Biochemistry

Agricultural Plant Biochemistry

Editor

Stefanos Palmer

scitus
academics

Agricultural Plant Biochemistry
Edited by **Stefanos Palmer**

ISBN: 978-1-68117-243-9
Library of Congress Control Number: 2016934783

www.scitusacademics.com
Box No. 4766, 616 Corporate Way,
Suite 2, Valley Cottage,
NY 10989

Printed in United States of America on Acid Free Paper

Preface

Plant Biochemistry is the study of chemical processes within and relating to living organisms. Plant Biochemistry is not only an important field of basic science explaining the molecular function of a plant, but is also an applied science that is in the position to contribute to the solution of agricultural and pharmaceutical problems.Plant biochemistry is an important emerging field in the agricultural sciences. Basic knowledge of the chemistry and the biochemical mechanisms of the plant in synthesizing various components are essential for advancements needed in other areas of agriculture like plant breeding, plant protection, plant production, etc. Plant Biochemistry is not only an important field of basic science explaining the molecular function of a plant, but is also an applied science that is in the position to contribute to the solution of agricultural and pharmaceutical problems. By controlling information flow through biochemical signaling and the flow of chemical energy through metabolism, biochemical processes give rise to the complexity of life. Over the last 40 years, biochemistry has become so successful at explaining living processes that now almost all areas of the life sciences from botany to medicine are engaged in biochemical research. Today, the main focus of Plant Biochemistry is in understanding how biological molecules give rise to the processes that occur within living cells, which in turn relates greatly to the study and understanding of whole organisms.The book, agricultural plant biochemistry, deals with the cellular and molecular biology and interaction between bimolecules along with the study of photosynthesis, respiration, plant nutrition, plant hormone functions which are associated with plant morphology, ecology and environmental effects on plants. The book is aimed at providing good information to graduate and post-graduate students in agriculture and biology. It will also serve as a valuable tool to researchers in plant breeding, agronomy, plant physiology and plant protection and will come in handy to solve many global problems by the present and future generations.

Table of Contents

1
CHAPTER

PROTEOMICS: A SUCCESSFUL APPROACH TO UNDERSTAND THE MOLECULAR MECHANISM OF PLANT-PATHOGEN INTERACTION

Tushar Dilipchand Lodha*, **Padmalochan Hembram***, **Nitile Tep, Jolly Basak**#

Department of Biotechnology, Visva-Bharati University, Santiniketan, India.

ABSTRACT

In recent years, proteomics has played a key role in identifying changes in protein levels in plant hosts upon infection by pathogenic organisms and in characterizing cellular and extracellular virulence and pathogenicity factors produced by pathogens. Proteomics offers a constantly evolving set of novel techniques to study all aspects of protein structure and function. Proteomics aims to find out the identity and amount of each and every protein present in a cell and actual function mediating specific cellular processes. Structural proteomics elucidates the development and application of experimental approaches to define the primary, secondary and tertiary structures of proteins, while functional proteomics refers to the development and application of global (proteome wide or system-wide) experimental approaches to assess protein function. A detail understanding of plant defense response using successful combination of proteomic techniques and other high throughput techniques of cell biology, biochemistry as well as genomics is needed for practical application to secure and stabilize yield of many crop plants. This review starts with a brief introduction to geland non gel-based proteomic techniques followed by the basics of plant-pathogen interaction, the use of proteomics in recent pasts to decipher the mysteries of plant-pathogen interaction, and ends with the future prospects of this technology.

Keywords: Proteome; Plant-Pathogen Interaction; Systemic Acquired Resistance; Mass Spectrometry; Pathogenicity Related Proteins

1. INTRODUCTION

Plant-pathogen interaction is a multifaceted process [1, 2]. At the beginning of the interaction, plants develop two pathways to recognize and resist pathogen attack. One pathway involves the formation of danger-associated molecular patterns (DAMPs) and pathogen associated molecular patterns (PAMPs) while other one results in effector-triggered immunity (ETI) and PAMPs-triggered immunity (PTI) after recognition by specific pathogen effector molecules [1-3]. Accordingly, downstream signaling cascades are activated, producing antimicrobial compounds that kill the pathogen and thereby maintain homeostasis [1-4]. Numerous proteins and signaling pathways are engaged in this precisely controlled multifaceted process [5]. This intricacy of plant-pathogen interactions makes it very difficult to determine which anatomical features, metabolites, and signaling pathways are triggered: conventional biochemical and genetic experimental methods are insufficient tools for the task [3]. Currently, proteomics provides a comprehensive insight to understand the intricacy of plant-pathogen interactions. Proteome is defined as the total protein components of the cell that are specified by the genome at specified condition and study of the structure and function of all these proteins present in a cell, organ or organism at a particular time is known as the proteomics [6]. Proteomics aims to find out the identity and amount of each and every protein present in a cell and actual function mediating specific cellular processes [7]. Structural proteomics elucidates the development and application of experimental approaches to define the primary, secondary and tertiary structures of proteins, while functional proteomics refers to the development and application of global (proteome wide or system-wide) experimental approaches to assess protein function.

Why we need proteomics at all in the successful post-genomic era?

1) In multicellular organism, although the DNA in each type of cell is same, different sets of cells express different sets of genes. Protein component of cell varies from cell to cell even under different stress conditions.

Therefore, by studying the proteome of individual cell, we can identify and analyze the proteins actually present therein.

2) Function of many genes identified by genome sequence remains a mystery [8].
3) Genome sequence tells us about the sequence of proteins, but there are many post translational modifications that are taking place in eukaryotic cells. Genomics fails to explain these modifications. Post translational modifications, the biological relevance of such modifications and transcript advances can only be interpreted through proteomics [9-12].

Modern techniques of proteomics allow us to study the protein content of cells which are differentially synthesized under diverse physiological and stress related conditions.

2. DIFFERENT TECHNIQUES USED IN PROTEOMICS

2.1. Two Dimensional Polyacrylamide Gel Electrophoresis (2D-PAGE)

2D-PAGE exploits the two independent properties (Isoelectric point pI and molecular size) of proteins to separate particular protein from the mixture of proteins [13,14]. Isoelectric focusing separates the proteins according to the charge present on them in first direction. For second direction, sodium dodecyl sulfate (SDS) PAGE is used to separate proteins according to their size [15]. 2D-PAGE provides a good platform to analyze the mixture of proteins. Under favorable circumstances, upto 5000 protein spots can be identified on a large format of 2D-gel. The separated protein spots are visualized by staining the gel with different staining techniques; Coomassie blue staining, silver staining or SYPRO staining [7,9]. SYPRO is a collection of molecular probe for protein detection. It stains for luminescent detection of proteins separated by PAGE. Digital images of 2D-PAGE are captured by sensitive scanners [9] and analyzed using softwares such as Melanie, PDQuest, Phoretix, Progenesis, Z3 or Z4000 [16]. The spots of interest are excised either manually or robotically from the gel [7]. After digestion with sequence specific proteases, the identity of proteins are analysed by mass spectrometry [17,18]. Analysis of 2D-PAGE gels simultaneously provides several pieces of informations such as molecular weight, isoelectric point, quantity as well as post translational modifications (PTMs) [15,19]. Various PTMs can be profiled via 2D-PAGE. Some specific applications are a) phosphor-protein expression profiling, based on the shift of pIs of phosphorylated protein to a more acidic region on the gel; b) acetylated-protein expression profiling based on the principle that acetylation of proteins at their amino termini or on lysine side-chains will cause them to shift to a more acidic region on the gel and (c) glycosylated-protein expression profiling based on the fact that attachment of relatively simple glycans (O-glycosylation) or more complex glycans (N-glycosylation) can have varying effects on protein molecular weight, pI, and protein functions.

2.2. Fluorescent Two-Dimensional "Difference Gel Electrophoresis" (2D-DIGE)

DIGE proteomics uses 2D gel electrophoresis to analyze differential protein regulation between control and target samples. In this technique, the two samples to be compared are each treated with one of two different but structurally similar fluorescent dyes (cy2, cy3 and cy5 etc.). Each dye reacts with amino groups, so that each protein is fluorescently labeled by the dye binding to lysine residues and the N-terminal amino group. The two protein mixtures to be compared are then mixed together and run on a single 2D gel. Thus every protein in one sample superimposes with its differentially labeled identical counterpart in other sample. Scanning of the gel at two different wavelengths that excite two dye molecules reveal whether any individual spot is associated with only one dye molecule rather than two [20]. The resulting images are then analysed by softwares which are specifically designed for 2D-DIGE analysis [21]. Advantages of DIGE is that it exhibits higher sensitivity as well as linearity, accuracy of quantification and enhanced reproducibility by directly comparing hundreds of protein samples under similar electrophoresis conditions [22, 23]. It allows rapid identification of protein changes between two samples on the same 2D gel without influence of gel to gel variation. Using DIGE technique, it is possible to compare hundred of proteins in a single experiment under quantitative and reproducible conditions. Since DIGE technique allows running an internal strand as part of each separation, more subtle abundance changes can be demonstrated with statistical confidence using the pooled internal standard methodology, thereby detecting true difference in protein expressions. Biomarker development involves reproducible, quantitative, highly sensitive and high-throughput experimental techniques, and 2D-DIGE meets these criteria, so it has also a great prospect in cancer proteomics toward personalized medicine.

Gel-based techniques, although very widely used, have several drawbacks including insensitivity to low copy proteins, poor reproducibility and inability to characterize the entire protein content. Because of the complexity of any given proteome and separation limits of 2D PAGE techniques, only fraction of the proteome can be analyzed [17]. An alternative approach is the gel-free proteomics technology, which is more suitable for the analysis of proteins with low abundance in complex samples. [7,17,24]. Separation of proteins using gel-free techniques revolutionized the study of proteome in the cell [24,25]. Several gel-free proteomic techniques are discussed below.

2.2.1. ICAT (Isotop-Coded Affinity Tags)

It is a gel-free method for quantitative proteomics that relies on chemical labeling reagents to identify and quantify most of the proteins present in two cell populations [7,10]. The isotope used in ICAT is C^{13}. Two populations are tagged with two different ICAT reagents (one with light and another with heavy), combined and enzymatically cleaved to generate peptide fragments, some of which are tagged. The tagged peptides are isolated by avidin affinity chromatography and subsequent analysis is done by microcapillary liquid chromatography-electron spray tandem mass spectrometry [7,10]. Combinations of the results are generated by MS and subsequent analysis of ICAT-labeled peptides determines the relative quantities of the components of protein mixtures in a single automated operation [10]. Gygi et al., demonstrated that they could follow the differential expression of more than 1400 different proteins in yeast [10]. This technique gives a quantitative as well as qualitative measure of changes in protein levels induced by different stress conditions. ICAT has been extensively used to find proteins associated with centrosomes (protein structures) that accumulate in abnormal ways in cancer cells and tumors. These proteins have the possibility to be used as diagnostic biomarkers and therapeutic targets (26).

2.2.2. ITRAQ (Isobarric Tagged for Relative and Absolute Quantitation)

ITRAQ is a non-gel based technique used to quantify protein from different sources in a single experiment [27-30]. Similar to ICAT, this technique also uses isotope coded covalent tags. The difference lies in the labeling of proteins. This method is based on covalent labeling of N-terminus and side chain amine of peptides from protein digestion with isotopic tag of varying mass. The peptides from the mixture are separated by nano chromatography. The separated peptides are then analyzed by tandem mass spectrometry (MS/MS). A database search is then performed using the fragmentation data to identify the labeled peptides and hence the corresponding proteins. The fragmentations of attached tag generate a low molecular mass reporter ion that can be used to relatively quantify the peptides and the proteins from which they originate using softwares (eg, i-tracker). The advantage of iTRAQ is its ability to multiplex several samples, quantification, simplified analysis and increased analytical precision and accuracy. It has the potential to comparing upto four different samples in one MS-based experiment.

2.2.3. MudPIT (Multidimensional Protein Identification Technology)

MudPIT is used especially for the analysis of complex protein mixtures [31]. In this technique, protein samples are subjected to sequence-specific enzymatic digestion with trypsin and endoproteinase lysc, and the resultant peptide mixtures are separated by strong cation exchange (SCX) and reverse phase (RP) high performance liquid chromatography (HPLC) [31,32]. Peptides from the RP column enter the mass spectrometer and MS data is used to search the protein databases [32]. The advantage of this technique is that it is fast and sensitive with good reproducibility and generates an exhaustive list of proteins present in a particular protein sample [18,32]. As MudPIT combines both SCX and RP-HPLC separation, it is successfully used in a broad range of proteomics experiments, including large-scale cataloging of proteins in cells and organisms, profiling of membrane and organelle proteins, identification of protein complexes, determination of PTMs and quantitative analysis of protein expression[33-35].

3. MASS SPECTROMETRY

Application of mass spectrometry to analyze amino acids and peptides was first reported in 1958 by Carl - Ove Anderson [36]. Mass spectrometry (MS) is an analytic technique which plays central role in the field of proteomics [37,38]. In MS, test samples are converted into gaseous ions that are separated in mass spectrometer according to their mass to charge ratio (m/z) and detected. In case of protein samples, proteins or peptides are fragmented by using the trypsin. The fragmented proteins are then separated using the liquid chromatography. The samples are then ionized and converted into the gaseous phase. Two soft ionization techniques, namely, Matrix-assisted laser desorption/ionization (MALDI) and Electron spray ionization (ESI) are used for the ionization. The invention of the time of flight mass spectrometry (TOF-MS) and relatively nondestructive methods to convert proteins into volatile ions made the revolution in the mass spectrometry [36]. Gaseous ions are separated in the mass analyzer. Peptides having a specific mass are fragmented using collision-induced dissociation and sent through a second mass spectrometry which generates a set of fragment peaks from which the amino acid sequence of the peptide are inferred. Protein identification algorithms are used to compare the results obtained with the known standards. Such algorithms fall into two categories; database search algorithms and de novo search algorithms. Some of the examples of the algorithms are listed in the Table 1. Mass spectrometry has a wide range of application in biotechnological development, pharmaceutical discovery, clinical testing as well as environmental analysis. Some of the common applications for MS are listed in the Table 2.

Table 1. List of the protein identification algorithms and data analysis softwares.

Program	Reference
Database search algorithms	
SEQUEST	[39]
Mascot	[40]
PEAKS Protein ID	[41]
Phenyx	[42]
OMSSA	[43]
MyriMatch	[44]
ByOnic	[45]
SIMS	[46]
MassWiz	[47]
***de novo* search programs**	
DeNoS	[48]
PEAKS	[49]
Other softwares	
ESIprot 1.0	[50]
Medicwave Bioinformatics Suite	www.medicwave.com
VIPER and Decon2LS	[51]
ProteoIQ	[52]
PatternLab for proteomics	[53]
MSight	[54]
Spectromania	[55]

Table 2. Applications and fields that uses mass spectrometry.

Field of Study	Applications
Proteomics	Determines protein structure, function, folding and interactions Identifies a protein from the mass of its peptide fragments Detects PTMs in complex biological mixtures Quantities proteins in a given sample
Drug Discovery	Determines structures of drugs and metabolites Screens metabolites in biological systems
Clinical Testing	Performs forensic analyses Detects disease biomarkers
Genomics	Sequences oligonucleotides
Environment	Test water quality or food contamination

4. PROTEIN MICROARRAY

Protein microarray is used for profiling complex samples in large quantities at a time and provides platform for both the classical and functional proteome analysis [56-60]. The first antibody microarray used for protein-protein interaction and post translational modification analysis in mammalian cells was reported in 2004 by Ivanov et al. [61]. Antibody array hold potential promise for the high-throughput profiling of a smaller number of proteins. In protein microarray antibodies or other affinity reagents such as short peptides, aptamers, polysaccharides, allergens or synthetic small molecules are spotted onto a surface such as glass, plastic or silicon chip; a complex mixture of cell lysate is passed over the surface to allow the antigen present to bind to their cognate antibodies. The screening of bound antigen is done by using fluorescently tagged or radioactively labeled proteins or by using secondary antibodies against each antigen of interest. Three types of protein microarray are currently used to study activity of protein. Analytical microarray is used to study differential expression profile, functional microarray is used for the study of protein-protein, protein-DNA, protein-RNA interaction and reverse phase protein microarray is used for determination of presence of altered protein that may be the result of disease.

5. PLANT-PATHOGEN INTERACTION

To gain an in depth understanding of plant-pathogen interactions, numerous studies have already been carried out from the plant as well as from the pathogen perspectives, which reveals that plant-pathogen interaction results from precise communication between the plant and the invading pathogen [1,3,60,62]. Plants serve as host for vast numbers of pathogens such as bacteria, fungi, nematodes, oomycetes, viruses, and insects [63]. In compatible interactions, plants are incapable of mounting effective anti-infectious defense responses, allowing the pathogens to complete their life cycle [3,63]. During incompatible interactions, plants trigger a series of complex defense responses against pathogenic interactions to forestall pathogen growth [3,60]. In general, two types of defenses are induced in plants during interaction with pathogens; local and systemic defense responses [3,60, 64]. Local response begins with recognition, when plant sense pathogens or pathogen isolates via pathogen associated molecular pattern (PAMPs) located on the cell surface (e.g. pattern recognition receptors) of plant [5]. This type of recognition is mediated by the plant resistance proteins and results in much more reliable, high impact defense responses, leading to rapid ion fluxes across the plasma membrane, cell wall reinforcement, MAP kinase activation, active production of reactive oxygen species (ROS) and culminating in programmed cell death (PCD) known as the hypersensitive response [5,60,64,65]. A systematic or long-term response not only restrict pathogen infection locally, but also induce signals that enhance defense responses to pathogens in distal systemic tissues, a phenomenon known as systemic acquired resistance (SAR). SAR plays crucial role in developing resistance to disease as well as to recover from the disease damage. Activation of SAR require endogenous salicylic acid (SA) and elevates the level of ethylene, jasmonic acid (JA), nitric oxide (NO), and pathogenesis related (PR) proteins, which in-turn activates many downstream processes [3,66-70]. The signaling pathways triggered by different stressors (e.g., pathogen attack and herbivores/wounding) have been widely documented [71]. Disentangling such complex defense systems require the investigation of plant reactions to different stressors in comparable conditions.

6. USE OF PROTEOMICS TO STUDY PLANT-PATHOGEN INTERACTIONS

Proteomics of plant-pathogen interactions started almost two decades ago when electrophoretic and protein identification techniques were not as high-tech as of today. Differential protein profiling of seedlings of Pinus lambertiana (sugar-pine), that were susceptible or resistant to the white pine blister rust fungus Cronartium ribicola, was carried out by Ekramoddoullah and Hunt in 1993 using 2D PAGE [72]. Foliage proteins were extracted and separated by 2D gel electrophoresis from three and nine days old pine seedlings after inoculating them with the C. ribicola. Two acidic proteins were detected in relatively large amounts, of which the bigger one (36.7 kDa) was significantly suppressed in susceptible seedlings at day three while the smaller one (28.1 kDa) was significantly enhanced in resistant seedlings at day nine. The differential expression of these two proteins accounts for the nature of the outcome of the infection process in the susceptible and resistant seedlings. Additionally, an increase of few proteins involved in protein biosynthesis and a decrease of others involved in the same process was observed in resistant seedlings whereas proteins involved in this vital cellular process were decreased only in susceptible seedlings [72].

Medicago truncatula is considered as the model plant for proteomic studies due to its small genome size (454- 526 Mbp). Numerous papers are published throughout the globe on proteomic study of M. trancatula [73-81]. Mathesius et al. [82] used a proteomic approach to detect the M trancatula response to the N-acyl homoserine lactone (AHL) signals from pathogenic Pseudomonas aeruginosa. Compared to the AHL from nitrogen fixing bacteria Sinorhizobium meliloti, they established that one third of the responsive elements are different. Using proteomics it could be confirmed that plant can distinguish between concentrations and time of exposure of AHLs from closely similar structures. Mathesius et al. has reported that in response to AHL, plant produces AHL "signal mimic" substances which manipulate the bacterial behavior controlled by quorum sensing.

Erwinia chrysanthemi is the causative agent of soft rot disease in ornamental plants and vegetables [83]. The proteome analysis of the secretome (extracellular proteins) of uninduced E. chrysanthemi using 2D-PAGE and MS showed that there are 55 spots on the 2D-PAGE gel from which 25 were uniquely present. Cellulase (Cel5), protease (PrtA, B, C), flagelin (Fli C) and some intracellular proteins were the major proteins found in uniduced condition. On the other hand, 14 unique proteins were successfully identified from the secretome of culture induced by a heat soluble extract prepared from Chrysanthemum leaves. Majority of these proteins are pectin lyases, pectin acetylsterase, pectin methylsterase and polygalacturonase. From the analysis of mutants, authors showed that the type-I system of secretome is playing key role in secretion of proteases and type-II system is responsible for the secretion of the esterase and of the Avr-like protein AvrL. AvrL is homologous to the avirulence protein reported in Xanthomonas campestries which is the causative agent of black rot disease in crucifers [84]. Watt et al., [85] did the proteome analysis of the X. campestries pv campestries using 2D PAGE and MALDI-TOF-MS. They found 97 distinct protein spots on coomassie brilliant blue stained gels, which were excised and tryptic digested fragments were analysed by MALDI-TOF-MS and 68 different proteins were identified. By Signal P software, authors concluded that only 53% of proteins have putative secretion signals which are secreted in sec-dependent pathway. Results showed that many of these extracellular proteins are involved in degradative activities and important factors allowing for the infection of susceptible plant hosts [85].

Zhang et al., [86] analyzed the accumulation of PR proteins induced by compatible and incompatible interactions of pepper mild mottle virus (PMMoV) in Capsicum chinense L3 plants using proteomics. Spanish strain (PMMoV-S) results in incompatible interaction while the Italian strain (PMMoV-I) results in compatible interaction, although both viruses are nearly identical at their nucleotide sequence level (98%). PMMoV-S induces a HR in C. chinense plants with the formation of necrotic local lesions and restriction of the virus at the primary infection sites [86]. 2D-PAGE and image analysis with PDQuest software detected differential accumulation of C. chinense PR proteins, of which isoforms belonging to the PR-1, β-1, 3-glucanases (PR-2), chitinases (PR-3), osmotin-like protein (PR-5), peroxidases (PR-9), germin-like protein (PR-16), and PRp27 (PR-17) were identified using MALDI-TOF and MS/MS and subsequent database search [68]. In HR condition, an earlier and higher accumulation of three of these isoforms, namely a basic PR-1, PR2 and PR5 were specifically induced during PMMoV-S-activation of C. chinense L3 plants. However, the accumulation rates of PR proteins did not correlated with maximal accumulation levels of viral RNA, indicating that PR protein expression reflects the physiological status of the plant [86].

Fagus sylvatica is a broad leaf species in Central and Eastern Europe and is highly susceptible to Phytophthora cambivora, P cinnamoni, P. citricola, P kernoviae, P gonapodyides, P. ramorum and P. cactorum. Out of these, P. citricola is particularly aggressive and the morphological and physiological consequences of F. sylvatica root infection with this pathogen have been addressed in several studies. The proteome analysis of the plant pathogen interaction between F. sylvatica and P. citrocola was studied by Valcu et al. in 2009. A total of 188 spots were identified in 2D-PAGE, out of which 142 were from leaf and 46 were from root. The synthesis of the proteins was changed following root infection with P. citricola or wounding. Of them, 129 proteins spots were regulated by pathogen attack, 37 spots specifically responded to wounding and 22 were common to both stresses. Protein spots were analysed by tandem MS and LC-MS/MS and was found that some proteins were up-regulated in infection and down regulated in wounding. The differentially expressed proteins were mostly translation elongation factors, protein kinases and mRNA binding translocases responsible for regulating general metabolism as well as synthesis of defense related proteins. Changes in protein abundance following root infection suggested a rather weak and inefficient defense against the pathogen [87]. Adjustments of host energy and primary metabolism consisted mainly in decreased carbon fixation accompanied by a reallocation of nutrients. Defense reactions against the pathogen were mobilized late and consisted of rather unspecific mechanisms involving phenylpropanoid and flavonoid synthesis pathways and protection from oxidative stress. The observed patterns of protein regulation suggested either a lack of recognition or a suppression of host's defense reaction by the invading pathogen, mimicking a sink organ and consequently manipulating host's metabolism to support its own growth [87].

Opium poppy (Papaver somniferum) is the source of pharmacologically important compounds. Many of these compounds are the result of defence response of poppy plants. Zulak et al., [88] used cell culture of opium poppy to study proteome of plant in induced condition by fungal-derived-elicitors. Proteome analysis was performed for the 50h elicitor-treated opium poppy cell cultures by 2D-PAGE coupled with LC-MS. Out of 340 spots on gel, 219 proteins were identified [88]. Within these 219 protein spots, 129 were identified in the pre-existing plant proteome databases, 63 were identified by comparing with predicted translation products in opium poppy EST databases, and the remaining were common in both databases [88,89]. Of the several categories, 38.4% of identified proteins were involved in metabolic processes representing the core category of proteins including several glycolytic and secondary metabolic enzymes, nearly entire set of TCA cycle enzymes, and one alkaloid biosynthetic protein [88]. The other categories of proteins were heat shock proteins, protein degradation factors, proteins involved in secondary metabolic and pathogenesis-related proteins providing a comprehensive proteomics view on the coordination of plant defense responses [88]. Secondary metabolic enzymes included a catechol Omethyltransferase, phenylalanine ammonia lyase, an Nmethyltransferase, three proteins similar to dirigent (DIR) proteins and 6OMT which converts (S)-norcoclaurine to (S)-coclaurine in benzylisoquinoline alkaloid biosynthesis. Six alkaloid biosynthetic enzymes were identified using available polyclonal antibodies which were absent in non induced culture [88]. The identification of proteins reported in this study is an aid to characterize the induction of benzylisoquinoline alkaloid metabolism and other defense responses in elicitor-treated opium poppy cell cultures, a so called non-model but important medicinal plant.

Table 3. A summary of important defense-related experiments using proteomics.

Plant species	Pathogen	No. of proteins identified	Proteomic technique used	Reference
Arabidopsis thaliana	*Alternaria brassicicola*	11	2D-PAGE, LC MS/MS	[97]
A. thaliana	*Spodoptera exigua*	19	2D-PAGE, MALDI-TOF	[95]
A. thaliana	*Pseudomonas syringe* pv. Tomato DC3000	12	Nitration of proteins, 2D-PAGE, LC MS/MS	[98]
A. thaliana	*Fusarium elicitor*	45	2D-DIGE, MALDI-TOF MS	[99]
A. thaliana	*Pseudomonas syringae*	52	2D-PAGE, LC-MS/MS	[100]
Brassica napus	*Sclerotinia sclerotiorum*	32	2D-PAGE, ESI-q-TOF MS/MS	[101]
B. napus & B. carinata	*Leptosphaeria maculans*	51	2D-PAGE, ESI-MS	[102]
Capsicum annuum	*Capsicum annuum*	9	2D-PAGE, MALDI-TOF/TOF-MS	[103]
Cucumis sativa	*Tricoderma asperellum*	28	2DE PAGE, MALDI-TOF/TOF	[104]
Coffea canephora	*Meloidogyne paranaensis*	2	2DE	[105]
Citrus sinensis	*X. axonopodis pv. citri*	5	2D-PAGE, N-terminal sequencing	[106]
Capsicum annuum	Tobacco mosaic virus	6	2D-PAGE, MALDI-TOF MS.	[89]
Fagus sylvatica	*Phytophthora citricola*	68	SDS-PAGE, 2D-PAGE, ESI, LC-ESI,	[87]
Glycine max	*Bacillus amyloliquefaciens*	20	2D PAGE	[107]
G. max	*Xanthomonas axonopodis pv. Glycine*	20	2D-PAGE	[107]
G. max	*Heterodera glycine*	28	2D-PAGE, q-TOF MS/MS	[108]
Gossipium hirsutum	*Meloidogyne Paranaensis*	1	2D-PAGE	[105]
Nicotiana benthamiana	PMMoV-S	2	2D-PAGE, N-terminal sequencing	[109]
L. hirsutum	*Clavibacter michiganensis* sp. *michiganensis*	47	2D-PAGE, LCQ-Deca ion trap MS	[110]
L. esculentum	*F. oxysporum*	21	2D-PAGE, MALDI-TOF MS, LC-QTOF MS/MS	[111]
Medicago truncatula	*Aphanomyces euteiches*	12	2D-PAGE, MALDI, PMF	[80]
M. truncatula	*Orobranche crenata*	49	2D- PAGE, PMF, MALDI-TOF	[112]
N. tabacum	pepper mild mottle virus	5	2D-PAGE, N-terminal sequencing	[109]
Oryza sativa	Rice yellow mottle virus (RYMV)	3	Exclusion chromatography, SDS-PAGE,Nano LC-MS/MS,	[113]
O. sativa	RYMV	32	2D-PAGE, MALDI, LC-MS/MS	[114]
O. sativa	*Nilaparvata lugens*	189	2D-PAGE,HPLC-MS/MS	[115]
O. sativa	*Rhizoctonia solani*	17	2D-PAGE, ESI Q-TOF MS, MS/MS	[89]
O. sativa	*Magnaporthe grisea*	8	2D-PAGE,MALDI-TOF,	[116]
O. sativa	*X. oryzae pv. oryzae*	20	2D-PAGE, MALDI TOF, Edman degradation	[117]
Olea europaea sub sp *Europaea*	*Pseudomonas savastanoi* pv. *Savastanoi*	7	2D-PAGE, MALDI TOF	[118]
Passiflora edulis	*Xanthomonas axonopodis* pv. *Passiflorae*	4	2D- PAGE, Edman sequencing	[119]
Prunus persica,and P. serotina	*Plum pox virus*	4	SDS-PAGE,IEF,MALDI-TOF	[120]
Solanum lycopersicum	*Ralstonia solanacearum*	12	2D- PAGE, n and cLC de novo	[121]
Saintpaulia ionantha	*Dickeya dadantii (syn. E. chrysanthemi)*	41	2D-PAGE, MALDI-TOF MS	[122]
Triticum aestivum	*Puccinia triticina*	30	2D-PAGE, MALDI-QqTOF MS/MS	[123]
T. aestivum	*Fusarium graminearum*	41	2D-PAGE, LC-MS/MS	[124]
Zea mays	*F. verticillioides*	16	2D-PAGE, MALDI-TOF-MS, LCQ ion trap MS	[125]
Vigna mungo	*Mungbean Yellow Mosaic Virus*	29	MALDI-TOF/TOF	[126]

Phaseolus vulgaris is a staple edible dry bean, having rust specific resistance to Uromyces appendiculatus. Lee et al., [89] studied the proteomics of P. vulgaris and U. appendiculatus interaction. Leaves were inoculated with liquid suspension of uredeospores of U. appendiculatus race 41 and 49. After 24 h and 72 h, the proteins were extracted and separated by differential centrifugation into three fractions; proteins from crude cell wall, organelle/membrane and soluble subcellular fraction. The fractions were analysed by LC-MS/MS. 1400 proteins were identified in each treatment and more than 3000 proteins were identified overall. The identified proteins obtained after 24hpi were categorized into defence (11%), energy (16%), metabolism (19%), primary destination and storage (14%), and protein synthesis (15%). At 72hpi the percentage was slightly less for defence (7%) and metabolism (14%) and slightly more for energy (22%) and protein destination and storage (17%). Lee et al., [89] assumed that similar type of early basal defense responses were functioning in both susceptible and resistant plants at 24h. Four proteins were identified with increased accumulation in both susceptible and resistant plants at 24 hpi including PR proteins, an inhibitor of bean rust differentiation and chitinase which is detrimental to chitin based cell walls of rust germlings. These results confirmed author's assumptions that basal defences were triggered in susceptible and resistant plant at 24 h. Lipolytic enzymes with a GDSL motif were increased in amount which triggers downstream defence responses. In these experiment it was found that the MLO1-like membrane spanning protein was increased and previously MLO1 like membrane spanning proteins were reported as a negative regulator of defence and cell death [89,90]. In late basal defence responses glutathione-S-transferases, O-methyltransferase, Cyclases and isopentenyl pyrophosphate isomerases, flavanoids, phytoalexins and other proteins that probably mobilize cellular toxins were increased. Lee et al. found that susceptible beans have equivalent potential to generate a strong resistance response but failed when infected with U. appendiculatus race 41, which likely impairs the innate basal resistance. Another new observation from this work was the fluctuation of proteins with roles in regulating Fatty Acid signaling, which is a major consequence of rust infection and plays important role in rust resistance.

The analysis of inducible defence mechanism in insect resistant plant can be effectively studied in Arabidopsis thaliana and Plutella xylostella interaction [91]. Collins et al., [92] carried out proteomic analysis of A. thaliana genotypes exhibiting resistance or susceptibility to the insect herbivore, P. xylostella. The extracted proteins were separated using 2D-PAGE and gel comparisons were performed using PDQuest. Protein identification was carried out using MASCOT search engine. It was found that 50 spots were abundant in the resistant recombinant inbreed lines (RILs), out of which 19 were identified and 17 spots were abundant in susceptible RILs from which 10 were identified. In resistant RILs, several proteins were unregulated, some of which have antioxidant role and were previously identified [93,94]. It was found that the P. xylostella larvae were lower in plants having increased levels of reactive oxygen species (ROS).

ROS has direct toxicity on the insect gut and upregulating defence genes [92]. Collins et al., concluded that significant differences exists in the proteomes between the identified resistant and susceptible RILs, and anticipated that enhanced production of ROS may be a major pre-existing mechanism of Plutella resistance in Arabidopsis Phytophthora sojae, a facultative pathogen of soybean, causes soybean root and stem rot, and results in an annual loss of 1 - 2 billion US dollars in soybean production worldwide. Zhang et al., [95] carried out proteomic study to determine the effects of P. sojae infection on soybean hypocotyls. Using 2D-PAGE and MALDI-TOF/ TOF, Zhang et al., identified 46 differentially expressed proteins. The expression levels of 26 proteins were significantly affected at various time points in the tolerant soybean line "Yudou25" (12 up-regulated and 14 downregulated). In contrast, in the sensitive soybean line "NG6255", only 20 proteins were significantly affected (11 up-regulated and 9 down-regulated). Among these proteins, 26% were related to energy regulation, 15% to protein destination and storage, 11% to defense against disease, 11% to metabolism, 9% to protein synthesis, 4% to secondary metabolism, and 24% were of unknown function [95]. The result from this study offers an insight into the repertoire of mechanisms used by P. sojae during its infection and colonization into a host.

Non-host resistance (NHR), a rarely studied defence phenomenon, is very complex due to the fact that activation of any specific defense component may not be sufficient to render a plant resistance reaction. Proteomics studies play a key role in delineating the molecular mechanism of non-host resistance. Li et al., [96] studied the NHR of Oryza sativa cv Nipponbare against the wheat leaf rust fungus, Puccinia triticina f. sp. tritici (Ptt). Proteins were extracted from leaves 3 dpi and analyzed by 2D-PAGE. A total of 33 protein spots were up-regulated and 9 were down-regulated by infection compared to the control, of which 30 were identified by MALDI-TOF-MS. The identified proteins were involved in defense/stress responses, energy/carbohydrate metabolism, oxidation-reduction processes, protein folding/turnover/cleavage/degradation, signal transduction and cell death regulation [96]. Proteomic analysis indicated that NHR of rice to Ptt is directly correlated with protein and energy metabolism, increased antimicrobial activities, possibly including phytoalexin accumulation and cell wall reinforcement, increased cell repair, antioxidative and detoxification reactions, and enhanced prevention of plant cell death [96]. Almost half of the up-regulated identified proteins were associated with chloroplast and mitochondrial physiology which suggests important roles for these organelles during NHR [96].

Till date, hundreds of proteomic experiments have been successfully carried out in the area of molecular plantmicrobe interaction, focusing mainly on the mechanisms controlling plant disease resistance, crosstalk among the signaling pathways involved, and the strategies used by the pathogens to suppress the defense. However, as space is limited, it is not possible to discuss all these interactions; Table 3 gives a summary of the important experiments involving plant-pathogen interaction using proteomics.

7. FUTURE PERSPECTIVES

With the experiments of plant-pathogen interaction completed to date, we can truly consider proteomics as a mature platform for proteome analysis during plant-pathogen interaction. Proteome wide functional classification using bioinformatics approaches is becoming an important method for revealing unknown protein functions. Proteomics enables scientists to decipher the functions of proteins and get a clearer picture of the complex regulatory networks that control fundamental biological processes [127]. To meet the current challenges of food insecurity, proteins that control crop architecture and/or stress resistance in a wide range of environments need to be identified to facilitate the biological improvement of crop productivity. To identify such proteins, state-ofthe-art proteomic techniques are the best choice of scientists. However, certain disadvantages are limiting the use of proteomics. Proteins are dynamic and interacting molecules, and their changeability can make proteomic snapshots difficult. Furthermore, very closely related proteins may not guarantee a functional relationship [128]. Accordingly, there is the need for more sensitive analytical systems and effective methods for large scale data comparison in order to advance our knowledge on protein expression during plant-pathogen interactions. Nevertheless, proteomic tools are rapidly improving and novel methods and equipments are being developed. We believe that future proteomic studies, coupled with functional validation as well as bioinformatical analysis, may provide new insights into plant disease resistance and pathogenicity. Proteomics will remain to be one of the fastest growing areas in research and a symbiotic relationship between proteomics technology and systems biology approaches in the future will enable new questions to be addressed in the area of plant–pathogen interaction.

8. ACKNOWLEDGEMENTS

TDL and NT are thankful to the Department of Biotechnology, Govt. of India, for providing financial assistance. PH is thankful to University Grant Commission, Govt. of India, for providing financial assistance and JB is thankful to Visva-Bharati for necessary facilities.

REFERENCES

1. P. N. Dodds and J. P. Rathjen, "Plant Immunity: Towards an Integrated View of Plant-Pathogen Interactions," Nature Reviews Genetics, Vol. 11, No. 8, 2010, pp. 539-548.
2. P. J. De Wit, "How Plants Recognize Pathogens and Defend Themselves," Cellular and Molecular Life Sciences, Vol. 64, No. 21, 2007, pp. 2726-2732.
3. T. D. Lodha and J. Basak, "Plant-Pathogen Interaction: What Microarray Tells about It?" Molecular Biotechnoogy, Vol. 50, No. 1, 2012, pp. 87-97.
4. J. L. Dangl and J. D. G. Jones, "Plant Pathogens and Integrated Defense Responses to Infection," Nature, Vol. 411, No. 6839, 2001, pp. 826-833.
5. C. Zipfel, "Pattern-Recognition Receptors in Plant Innate Immunity," Current Opinion in Immunology, Vol. 20, No. 1, 2008, pp. 10-16.
6. M. R. Wilkins, J. C. Scnchez, A. A. Gooley, R. D. Appel, I. H. Smith, D. F. Hochstrasser and K. L. Williams, "Progress with Proteome Projects: Why All Proteins Expressed by a Genome Should Be Identified and How to Do It," Biotechnology & Genetic Engineering Reviews, Vol. 13, No. 1, 1995, pp. 19-50.
7. N. V. K. Nat, S. Srivastava, W. Yajima and N. Sharma, "Application of Proteomics to Investigate Plant-Pathogen Interactions," Current Proteomics, Vol. 4, No. 1, 2007, pp. 28-43.
8. S. D. Patterson and R. H. Aebersold, "Proteomics: The First Decade and Beyond," Nature Genetics, Vol. 33, No. 3, 2003, pp. 311-321.
9. H. Duley and A. Grover, "Current Initiatives in Proteomics Research: The Plant Perspective," Current Science, Vol. 80, No. 2, 2001, pp. 262-269.
10. S. P. Gygi, B. Rist, S. A. Gerber, F. Turecek, M. H. Gelb and R. Aebersold, "Quantitative Analysis of Complex Protein Mixtures Using Isotope-Coded Affinity Tags," Nature Biotechnology, Vol. 17, No. 10, 1999, pp. 994-999.

11. O. K. Park, "Proteomic Study in Plants," Journal of Biochemistry and Molecular Biology, Vol. 37, No. 1, 2004, pp. 133-138.
12. G. Thurston, S. Regan, C. Rampitsch and T. Xing, "Proteomic and Phosphoproteomic Approaches to Understand Plant-Pathogen Interactions," Physiological and Molecular Plant Pathology, Vol. 66, No. 1, 2005, pp. 3-11.
13. J. Klose, "Protein Mapping by Combined Isoelectric Focusing and Electrophoresis of Mouse Tissues. A Novel Approach to Testing for Induced Point Mutation in Mammals," Human Genetics, Vol. 26, No. 3, 1975, pp. 231- 243.
14. P. H. O'Farrell, "High Resolution Two Dimensional Electrophoresis of Proteins," Journal of Biological Chemistry, Vol. 250, No. 10, 1975, pp. 4007-4021.
15. A. Gorg, W. Weiss and M. J. Dunn, "Current Two-Dimensional Electrophoresis Technology for Proteomics," Proteomics, Vol. 4, No. 12, 2004, pp. 3665-3685.
16. P. G. Righetti, A. Castagna, F. Antonucci, C. Piubelli, D. Cecconi, N. C. Campostrini, P. Antonioli, H. Astner, "Critical Survey of Quantitative Proteomics in Two-Dimensional Electrophoretic Approaches," Journal of Chromatography A, Vol. 1051, No. 1, 2004, pp. 3-17.
17. H. Zhu, M. Bilgin and M. Snyder, "Proteomics," Annual Review of Biochemistry, Vol. 72, No. 1, 2003, pp. 783-812.
18. J. K. C. Rose, S. Bashir, J. J. Giovannoni, M. M. Jahn and R. S. Saravanan, "Tackling the Plant Proteome: Practical Approaches, Hurdles and Experimental Tools," Plant Journal, Vol. 39, No. 5, 2004, pp. 715-733.
19. B. Wittmann-Liebold, H. R. Graack and T. Pohl, "TwoDimensional Gel Electrophoresis as Tool for Proteomics Studies in Combination with Protein Identification by Mass Spectrometry," Proteomics, Vol. 6, No. 17, 2006, pp. 4688-4703.
20. M. Unlu, M. E. Morgan and J. S. Minden, "Difference Gel Electrophoresis: A Single Gel Method for Detecting Changes in Protein Extracts," Electrophoresis, Vol. 18, No. 11, 1997, pp. 2071-2077.
21. R. Marouga, S. David and E. Hawkins, "The Development of the DIGE System: 2D Fluorescence Difference Gel Analysis Technology," Analytical and Bioanalytical Chemistry, Vol. 382, No. 3, 2005, pp. 669-678.

22. G. Van den Bergh and L. Arckens, "Recent Advances in 2D Electrophoresis: An Array of Possibilities," Expert Review of Proteomics, Vol. 2, No. 2, 2005, pp. 243-252.
23. G. Zhou, H. Li and D. Camp De, "2D Differential In-Gel Electrophoresis for the Identification of Esophageal Scans Cell Cancer-Specific Protein Markers," Molecular & Cellular Proteomics, Vol. 1, No. 2, 2005, pp. 117-124.
24. J. P. Lambert, M. Ethier, J. C. Smith and D. Figeys, "Proteomics from Gel Based to Gel Free," Analytical Chemistry, Vol. 77, No. 12, 2005, pp. 3771-3787.
25. G. Baggerman, E. Vierstraete, A. De Loof and L. Schoofs, "Gel Based Versusgel-Free Proteomics: A Review," Combinatorial Chemistry & High Throughput Screening, Vol. 8, No. 8, 2005, pp. 669-677.
26. M. R. Flory, A. R. Carson, E. G. Muller and R. Aebersold, "An SMC-Domain Protein in Fission Yeast Links Telomeres to the Meiotic Centrosome," Molecular Cell, Vol. 16, No. 4, 2004, pp. 619-630.
27. P. L. Ross, Y. N. Huang, J. N. Marchese, B. Williamson, K. Parker, S. Hattan, N. Khainovski and S. Pillai, "Multiplexed Protein Quantitation in Saccharomyces cerevisiae Using Amine-Reactive Isobaric Tagging Reagents," Molecular & Cellular Proteomics, Vol. 3, No. 12, 2004, pp. 1154-1169.
28. K. Agarwal, L. H. Choe and K. H. Lee, "Shotgun Proteomics Using the iTRAQ Isobaric Tags," Briefings in Functional Genomics and Proteomics, Vol. 5, No. 2, 2006, pp. 112-120.
29. L. R. Zieske, "A Perspective on the Use of iTRAQ Reagent Technology for Protein Complex and Profiling Studies," Journal of Experimental Botany, Vol. 57, No. 7, 2006, pp. 1501-1508.
30. T. C. Lund, L. B. Anderson, V. McCullar, L. Higgins, G. H. Yun, B. Grzywacz, M. R. Verneris and J. S. Miller, "iTRAQ Is a Useful Method to Screen for MembraneBound Proteins Differentially Expressed in Human Natural Killer Cell Types," Journal of Proteome Research, Vol. 6, No. 2, 2007, pp. 644-653.
31. H. J. Issaq, K. C. Chan, G. M. Janini, T. P. Conrads and T. D. Veenstra, "Multidimensional Separation of Peptides for Effective Proteomic Analysis," Journal of Chromatography B Analytical Technologies in the Biomedical and Life Sciences, Vol. 817, No. 1, 2005, pp. 35-47.

32. M. P. Washburn, D. Wolters and J. R. Yates, "LargeScale Analysis of the Yeast Proteome by Multidimensional Protein Identification Technology," Nature Biotechnology, Vol. 19, No. 3, 2001, pp. 242-247.
33. I. I. I. Yates, J. R. A. Gilchrist, K. E. Howell and J. J. Bergeron, "Proteomics of Organelles and Large Cellular Structures," Nature Reviews, Vol. 6, No. 9, 2005, pp. 702-714.
34. A. E. Speers and C. C. Wu, "Proteomics of Integral Membrane Proteins-Theory and Application," Chemical Reviews, Vol. 107, No. 8, 2007, pp. 3687-3714.
35. G. T. Cantin, J. D. Venable, D. Cociorva and J. R. Yates III, "Quantitative Phosphoproteomic Analysis of the Tumor Necrosis Factor Pathway," Journal of Proteome Research, Vol. 5, No. 1, 2006, pp. 127-134.
36. C .O. Anderson, "Mass Spectrometric Studies on Amino Acid and Peptide Derivatives," Acta Chemica Scandinavica, Vol. 12, No. 0353, 1958, p. 1353.
37. W. Zhu, J. W. Smith and C. M. Huang, "Mass Spectrometry-Based Label-Free Quantitative Proteomics," Journal of Biomedicine and Biotechnology, Vol. 2010, 2009, Article ID: 840581.
38. M. Zhu, B. Simons, N. Zhu, D. G. Oppenheimer and S. Chen, "Analysis of Abscisic Acid Responsive Proteins in Brassica Napus Guard Cells by Multiplexed Isobaric Tagging," Journal of Proteomics, Vol. 73, No. 4, 2010, pp. 790-805.
39. J. K. Eng, L. A. McCormack and J. R. Yates, "An Approach to Correlate Tandem Mass Spectral Data of Peptides with Amino Acid Sequences in a Protein Database," Journal of the American Society for Mass Spectrometry, Vol. 5, No. 11, 1994, pp. 976-989.
40. D. N. Perkins, D. J. C. Pappin, D. M. Creasy and J. S. Cottrell, "Probability-Based Protein Identification by Searching Sequence Databases Using Mass Spectrometry Data," Electrophoresis, Vol. 20, No. 18, 1999, pp. 3551- 3567.
41. C. Liang, J. C. Smith and C. Hendrie, "A Comparative Study of Peptide Sequencing Software Tools for MS/ MS," Proceedings of the 51st ASMS Conference on Mass Spectrometry and Allied Topics, Montreal, 8-13 June 2003.
42. C. Jacques, A. Masselot, M. Giron, T. Dessingy and J. Magnin, "OLAV: Towards High-Throughput Tandem Mass Spectrometry Data Identification," Proteomics, Vol. 3, No. 8, 2003, pp. 1454-1463.

43. Y. G. Lewis, S. P. Markey, J. A. Kowalak, L. Wagner, M. Xu, D. M. Maynard, X. Yang and W. Shi, "Open Mass Spectrometry Search Algorithm," Journal of Proteome Research, Vol. 3, No. 5, 2004, pp. 958-964.
44. L. T. David, C. G. Fernando and M. C. Chambers, "MyriMatch: Highly Accurate Tandem Mass Spectral Peptide Identification by Multivariate Hypergeometric Analysis," Journal of Proteome Research, Vol. 6, No. 2, 2007, pp. 654-661.
45. B. Marshall, Y. Cai and D. Goldberg, "Lookup Peaks: A Hybrid of de Novo Sequencing and Database Search for Protein Identification by Tandem Mass Spectrometry," Analytical Chemistry, Vol. 79, No. 4, 2007, pp. 1393- 1400.
46. L. Jian, A. Erassov, P. Halina, M. Canete, N. D. Vo, C. Chung, G. Cagney and A. Ignatchenko, "Sequential Interval Motif Search: Unrestricted Database Surveys of Global MS/MS Data Sets for Detection of Putative Post-Translational Modifications," Analytical Chemistry, Vol. 80, No. 20, 2008, pp. 7846-7854.
47. A. K. Yadav, D. Kumar and D. Dash, "MassWiz: A Novel Scoring Algorithm with Target-Decoy Based Analysis Pipeline for Tandem Mass Spectrometry," Journal of Proteome Research, Vol. 10, No 5, 2011, pp. 2154-2160.
48. M. S. Mikhail, M. L. Nielsen, F. Kjeldsen and R. A. Zubarev, "Proteomics-Grade de Novo Sequencing Approach," Journal of Proteome Research, Vol. 4, No. 6, 2005, pp. 2348-2354.
49. M. Bin, K. Zhang, C. Hendrie, C. Liang, M. Li, A. Doherty-Kirby and G. Lajoie, "PEAKS: Powerful Software for Peptidede Novo Sequencing by Tandem Mass Spectrometry," Rapid Communication in Mass Spectrometry, Vol. 17, No. 20, 2003, pp. 2337-2342.
50. R. Winkler, "ESIprot: A Universal Tool for Charge State Determination and Molecular Weight Calculation of Proteins from Electrospray Ionization Mass Spectrometry Data," Rapid Communication in Mass Spectrometry, Vol. 24, No. 3, 2010, pp. 285-294.
51. M. E. Monroe, N. Tolic, N. Jaitly, J. L. Shaw, J. N. Adkins and R. D. Smith, "VIPER: An Advanced Software Package to Support High-Throughput LC-MS Peptide Identification," Bioinformatics, Vol. 23, No. 15, 2007, pp. 2021-2023.
52. D. B. Weatherly, J. A. Atood, T. A. Minning, C. Cavola, R. L. Tarleton and R. Orlando, "A Heuristic Method for Assigning a False-Discovery Rate

for Protein Identifications from Mascot Database Search Results," Molecular and Cellular Proteomics, Vol. 4, No. 6, 2005, pp. 762- 772.

53. C. C. Paulo, J. S. G. Fischer, E. I. Chen, J. R. Yates and V. C. Barbosa, "PatternLab for Proteomics: A Tool for Differential Shotgun Proteomics," BMC Bioinformatics, Vol. 9, 2008, p. 316.
54. M. P. Patricia, D. Walther, M. Quadroni, S. B. Catherinet, J. Burgess, C. G. Zimmermann-Ivol, J. C. Sanchez and P. A. Binz, "MSight: An Image Analysis Software for Liquid Chromatography-Mass Spectrometry," Proteomics, Vol. 5, No. 9, 2005, pp. 2381-2384.
55. Z. Hans-Dieter, L. Jens., V. Khamenia, C. Schiller, A. Appel, H. Tammen, R. Crameri and H. Selle, "Datamining Methodology for LC-MALDI-MS Based Peptide Profiling," Combinatorial Chemistry and High Throughput Screening, Vol. 8, No. 8, 2005, pp. 717-723.
56. X. Duburcq, C. Olivier, F. Malingue, R. Desmet, A. Bouzidi, F. Zhou, C. Auriault and H. Gras-Masse, "Peptide-Protein Microarrays for the Simultaneous Detection of Pathogen Infections," Bioconjugate Chemistry, Vol. 15, No. 2, 2004, pp. 307-316.
57. J. LaBaer and N. Ramachandran, "Protein Microarrays as Tools for Functionalproteomics," Current Opinion in Chemical Biology, Vol. 9, No. 1, 2005, pp. 14-19.
58. G. S. Athwal, C. Lombardo, J. L. Huber, S. C. Masters, H. Fu and S. C. Huber, "Modulation of 14-3-3 Interactions with Target Proteins by Physical and Metabolic Effectors," Plant and Cell Physiology, Vol. 41, No. 4, 2000, pp. 523-533.
59. N. Ramachandran, D. N. Larson, P. R. Stark, E. Hainsworth and J. LaBaer, "Emerging Tools for Real-Time Label-Free Detection of Interactions on Functional Protein Microarrays," FEBS Journal, Vol. 272, No. 21, 2005, pp. 5412-5425.
60. K. E. Hammond-Kosack and J. D. G. Jones, "Responses to Plant Pathogens," In: B. B. Buchanan, W. Gruissem and R. L. Jones, Eds., Biochemistry and Molecular Biology of Plants, American Society of Plant Physiology, Rockville, 2000, pp. 1102-1156.
61. S. S. Ivanov., et al, "Antibodies Immobilized as Arrays to Profile Protein Post-Translational Modifications in Mammalian Cells," Molecular and Cellular Proteomics, Vol. 3, No. 8, 2004, pp. 788-795.
62. J. A. Lucas, "Plant Pathology and Plant Pathogens," Blackwell Science, Oxford, 1998.

63. P. M. Schenk, K. Kazan, L. Wilson, J. P. Anderson, T. Richmond and S. C. Somerville, "Coordinated Plant Defense Responses in Arabidopsis Revealed by Microarray Analysis," Proceedings of National Academy of Science, Vol. 97, No. 21, 2000, pp. 11655-11660.
64. P. M. Schenk, J. H. Choo and C. L. Wong, "Microarray Analyses to Study Plant Defense and Rhizosphere Microbe Interaction," CAB Review: Perspective in Agriculture, Veterinary, Science, Nutrition, and Natural Resources, Vol. 4, No. 045, 2009, pp. 1-14.
65. L. A. Mur, P. Kenton, A. J. Lloyd, H. Ougham and E. Prats, "The Hypersensitive Response; The Centenary Is upon Us but How Much Do We Know?" Journal of Experimental Botany, Vol. 59, No. 3, 2008, pp. 501-520.
66. F. B. Andrew, W. Jinrong and D. F. Mark, "Probing Plant-Pathogen Interaction and Downstream Defense Signaling Using DNA Microarray," Functional & Integrative Genomics, Vol. 2, No. 6, 2002, pp. 259-273.
67. M. A. Torres, D. G. Jonathan and J. L. Dangl, "Reactive Oxygen Species Signaling in Response to Pathogen," Plant Physiology, Vol. 141, No. 2, 2006, pp. 373-378.
68. G. E. Vallad and R. M. Goodman, "Systemic Acquired Resistance and Induced Systemic Resistance in Conventional Agriculture," Crop Science, Vol. 44, No. 6, 2004, pp. 1920-1934.
69. X. Dong, "Genetic Dissection of Systemic Acquired Resistance," Current Opinion in Plant Biology, Vol. 4, No. 4, 2001, pp. 309-314.
70. J. A. Kreps, Y. Wu, H. S. Chang, T. Zhu, X. Wang and J. F. Harper, "Transcriptome Changes for Arabidopsis in Response to Salt, Osmotic, and Cold Stress," Plant Physiology, Vol. 130, No. 4, 2002, pp. 2129-2141.
71. C. M. J. Pieterse and L. C. Van Loon, "NPR1: The Spider in the Web of Induced Resistance Signaling Pathways," Current Opinion in Plant Biology, Vol. 7, No. 4, 2004, pp. 456-464.
72. A. K. M. Ekramoddoullah and R. S. Hunt, "Changes in Protein Profile of Susceptible and Resistant Sugar-Pine Foliage Infected with the Whitepine Blister Rust Fungus Cronartium ribicola," Canadian Journal in Plant Pathology, Vol. 15, No. 4, 1993, pp. 259-264.
73. V. S. Asirvatham, B. S. Watson and L. W. Sumner, "Analytical and Biological Variances Associated with Proteomic Studies on Medicago Trancatula by Two Dimensional Polyacrylamide Gel Electrophoresis," Proteomics, Vol. 2, No. 8, 2002, pp. 960-968.

74. G. Bestel-Corre, E. Dumas-Gaudot, V. Poinsot, M. Dieu, J. F. Dierick, D. Van Tuinen, J. Remacle, V. Gianinazzipearson and S. Gianinazzi, "Proteome Analysis and Identification of Symbiosis-Related Proteins from Medicago truncatula Gaertn. By Two Dimensional Electrophoresis and Mass Spectrometry," Electrophoresis, Vol. 23, No. 1, 2002, pp. 122-137.
75. G. Bestel-Corre, S. Gianinazzi and E. Dumass-Gaudot, "Impact of Sewage Sludge on Medicago truncatula Symbiotic Proteome," Phytochemistry, Vol. 65, No. 11, 2004, pp. 1651-1659.
76. E. Dumas-Gaudot, N. Amiour, S. Weidmann, G. BestelCorre, B. Valot, S. Lenogue, V. Gianninazzi-Pearson and S. Gianninazzi, "A Technical Trick for Studying Proteomics in Parallel to Transcriptomics in Symbiotic RootFungus Interaction," Proteomics, Vol. 4, No. 2, 2004, pp. 451-453.
77. K. Gallardo, C. Le Signor, J. Vandekerckhove, R. D. Thompson and J. Burstin, "Proteomics of Medicago truncaula Seed Development Establishes the Time Fram of Diverse Metabolic Processes Related to Reverse Accumulation," Plant Physiology, Vol. 133, No. 2, 2003, pp. 664-682.
78. U. Mathesius, G. Keijzers, S. H. A. Natera, J. J. Weinman, M. A. Djordjeivic and B. G. Rolfe, "Establishment of a Root Proteome Reference Map for the Model Legume Medicago truncatula Using the Expressed Sequence Tag Database for Peptide Mass Fingerprinting," Proteomics, Vol. 1, No. 11, 2001, pp. 1424-1440.
79. B. S. Watson, Z. Lei, R. A. Dixon and L. W. Sumner, "Proteomics of Medicago sativa Cell Walls," Phytochemistry, Vol. 65, No. 12, 2004, pp. 1709-1720.
80. F. Colditz, O. Nyamsuren, K. Niehaus, H. Eubel, H. P. Braun and F. Krajinski, "Proteomic Approach: Identification of Medicago truncatula Proteins Induced in Roots after Infection with the Pathogenic Oomycete Aphanomyces euteiches," Plant Molecular Biology, Vol. 55, No. 1, 2004, pp. 109-120.
81. M. A. Castillejo, N. Amiour, E. Dumas-Gaudot, D. Rubiales and J. V. Jorrın, "A Proteomic Approach to Studying Plant Response to Crenatebroomrape (Orobanche crenata) in Pea (Pisum sativum)," Phytochemistry, Vol. 65, No. 12, 2004, pp. 1817-1828.
82. U. Mathesius, S. Mulders, M. S. Gao, M. Teplitski, G. Caetano-Anolles, B. G. Rolfe and W. D. Bauer, "Extensive and Specific Responses of a Eukaryote to Bacterial Quorum-Sensing and Signals," Proceedings of

National Academy of Science of the United States of America, Vol. 100, No. 3, 2003, pp. 1444-1449.
83. F. Bouchart, A. Delangle, J. Lemoine, J. P. Bohin and J. M. Lacroix, "Proteomic Analysis of a Non-Virulent Mutant of the Phytopathogenic Bacterium Erwinia chrysanthemi Deficient in Osmoregulated Periplasmic Glucans: Change in Protein Expression is Not Restricted to the Envelope, but Affects General Metabolism," Microbiology, Vol. 153, No. 3, 2007, pp. 760-767.
84. N. Kazemi-Pour, G. Condemine and N. HugouvieuxCotte-Pattat, "The Secretome of the Plant Pathogenic Bacterium Erwinia chrysanthemi," Proteomics, Vol. 4, No. 10, 2004, pp. 3177-3186.
85. S. A. Watt, A. Wilke, T. Patschkowski and K. Niehaus, "Comprehensive Analysis of the Extracellular Proteins from Xanthomonas campestris pv. campestris B100," Proteomics, Vol. 5, No. 1, 2005, pp. 153-167.
86. M. I. Elvira, M. M. Galdeano, P. Gilardi, I. Garcı´a-Luque, M. T. Serra, "Proteomic Analysis of PathogenesisRelated Proteins (PRs) Induced by Compatible and Incompatible Interactions of Pepper Mild Mottle Virus (PMMoV) in Capsicum chinense L^3 Plants," Journal of Experimental Botany, Vol. 59, No. 6, 2008, pp. 1253- 1265.
87. C. M. Valcu, M. Junqueira, A. Shevchenko and K. Schlink, (2009) Comparative Proteomic Analysis of Responses to Pathogen Infection and Wounding in Fagus sylvatica," Journal of Proteome Research, Vol. 8, No. 8, 2009, pp. 4077-4091.
88. K. G. Zulak, F. K. Morgan, A. Joenel, C. S. David and J. F. Peter, "Plant Defense Responses in Opium Poppy Cell Cultures Revealed by Liquid Chromatography-Tandem Mass Spectrometry Proteomics,' Molecular & Cellular Proteomics, Vol. 8, No. 1, 2009, pp. 86-98.
89. J. Lee, J. Feng, K. B. Campbell, B. E. Scheffler, W. M. Garrett, S. Thibivilliers, G. Stacey, D. Q. Naiman, M. L. Tucker, M. A. Pastor-Corrales and B. Cooper , "Quantitative Proteomic Analysis of Bean Plants Infected by a Virulent and Avirulent Obligate Rust Fungus," Molecular & Cellular Proteomics, Vol. 8, 2009, pp. 19-31.
90. P. Piffanelli, F. Zhou, C. Casais, J. Orme, B. Jarosch, U. Schaffrath, N. C. Collins, R. Panstruga and P. SchulzeLefert, "The Barley MLO Modulator of Defense and Cell Death Is Responsive to Biotic and Abiotic Stress Stimuli," Plant Physiology, Vol. 129, No. 3, 2002, pp. 1076- 1085.
91. D. Kliebenstein, D. Pedersen, B. Barker and T. MitchellOlds, "Comparative Analysis of Quantitative Trait Loci Controlling

Glucosinolates, Myrosinase and Insect Resistance in Arabidopsis thaliana," Genetics, Vol. 161, No. 1, 2002, pp. 325-332.

92. R. M. Collins, M. Afzal, D. A. Ward, M. C. Prescott, S. M. Sait, H. H. Rees and A. D. Tomsett, "Differential Proteomic Analysis of Arabidopsis thaliana Genotypes Exhibiting Resistance or Susceptibility to the Insect Herbivore, Plutella xylostella," PLoS One, Vol. 5, No. 4, 2010, e10103.
93. G. P. Biewenga, G. R. M. M. Haenen and A. Bast, "The Pharmacology of the Antioxidant Lipoic Acid," General Pharmacology: The Vascular System, Vol. 29, No. 3, 1997, pp. 315-331.
94. A. Koshkin, G. M. Knudsen and P. R. O. de Montellano, "Intermolecular Interactions in the AhpC/AhpD Antioxidant Defence System of Mycobacterium tuberculosis," Archives of Biochemistry and Biophysics, Vol. 427, No. 1, 2004, pp. 41-47.
95. J. H. Zhang, L. W. Sun, L. L. Liu, S. L. An, X. Wang, J. Zhang, J. L. Jin, S. Y. Li and J. H. Xi, "Proteomic Analysis of Interactions between the Generalist Herbivore Spodoptera exigua (Lepidoptera Noctuidae) and Arabidopsis thaliana," Plant Molecular Biology Reporter, Vol. 28, No. 2, 2010, pp. 324-333.
96. H. Li, P. H. Goodwin, Q. Han, L. Huang and Z. Kang, "Microscopy and Proteomic Analysis of the Non-Host Resistance of Oryza sativa to the Wheat Leaf Rust Fungus, Puccinia triticina f. sp. tritici," Plant Cell Reports, Vol. 31, No. 4, 2012.
97. A. K. Mukherjee, M. J. Carp, R. Zuchman, T. Ziv, B. A. Horwitz and S. Gepstein, "Proteomics of the Response of Arabidopsis thaliana to Infection with Alternaria brassicicola," Journal of Proteomics, Vol. 73, No. 4, 2010, pp. 709-720.
98. D. Cecconi, S. Orzetti, E. Vandelle, S. Rinalducci, L. Zolla and M. Delledonne, "Protein Nitration during Defese Response in Arabidopsis thaliana," Electrophoresis, Vol. 30, No. 14, 2009, pp. 2460-2468.
99. S. Chivasa, J. M. Hamilton, R. S. Pringle, B. K. Ndimba, W. J. Simon, K. Lindsey and A. R. Slabas, "Proteomic Analysis of Differentially Expressed Proteins in Fungal Elicitor-Treated Arabidopsis Cell Cultures," Journal of Experimental Botany, Vol. 57, No. 7, 2006, pp. 1553- 1562.
100. A. M. E. Jones, V. Thomas, M. H. Bennett, J. Mansfield, and M. Grant, "Modifications to the Arabidopsis Defence Proteome Occur Prior to Significant Transcriptional Change in Response to Inoculation with

Pseudomonas syringae," Plant Physiology, Vol. 142, No. 4, 2006, pp. 1603-1620.

101. Y. Liang, S. Srivastava, M. H. Rahman, S. E. Strelkov, and N. N. Kav, "Proteome Changes in Leaves of Brassica napus L. as a Result of Sclerotinia sclerotiorum Challenge," Journal of Agricultural and Food Chemistry, Vol. 56, No. 60, 2008, pp. 1963-1976.

102. N. Sharma, N. Hotte, M. H. Rahman, M. Mohammadi, M. K. Deyholos and N. N. Kav, "Towards Identifying Brassica Proteins Involved in Mediating Resistance to Leptosphaeria maculans: A Proteomics-Based Approach," Proteomics, Vol. 8, No. 17, 2008, pp. 3516-3535.

103. A. Wongpia and K. Lomthaisong "Changes in the 2DE Protein Profiles of Chilli Pepper," ScienceAsia, Vol. 36, 2010, pp. 259-270.

104. G. Segarra, E. Casanova, D. Bellido, M. A. Odena, E. Oliveira and I. Trillas, "Proteome, Salicylic Acid and Jasmonic Acid Changes in Cucumber Plants Inoculated with Trichoderma asperellum Strain T34," Proteomics, Vol. 7, No. 21, 2007, pp. 3942-3952.

105. O. L. Franco, J. L. Pereira, P. H. A. Costa, T. L. Rocha, E. V. S. Albuquerque, M. F. Grossi-de-Sa, R. M. D. G. Carneiro, R. G. Carneiro and A. Mehta, "Methodological Evaluation of 2DE to Study Root Proteomics during Nematode Infection in Cotton and Coffee Plants," Preparative Biochemistry and Biotechnology, Vol. 40, No. 2, 2010, pp. 152-163.

106. A. Mehta and Y. B. Rosato, "Differentially Expressed Proteins in the Interaction of Xanthomonas axonopodis pv. citri with Leaf Extract of the Host Plant," Proteomics, Vol. 1, No. 9, 2001, pp. 1111-1118.

107. N. Buensanteai, G. Y. Yuen and S. Prathuangwong, "Priming, Signaling and Protein Production Associated with Induced Resistance by Bacillus amyloliquefaciens KPS46," World Journal of Microbiology and Biotechnology, Vol. 25, No. 7, 2009, pp. 1275-1286.

108. A. J. Afzal, A. Natarajan, N. Saini, M. J. Iqbal, M. Geisler, H. A. El-Shemy, R. Mungur, L. Willmitzer and D. A. Lightfoot, "The Nematode Resistance Allele at the rhg1 Locus Alters the Proteome and Primary Metabolism of Soyabean Roots," Plant Physiology, Vol. 151, No. 3, 2009, pp. 1264-1280.

109. M. L. Perez-Bueno, J. Rahoutei, C. Sajnani, I. Garcia-Luque and M. Baron, "Proteomic Analysis of the Oxygen-Evolving Complex of Photosystem II under Biotic Stress: Studies on Nicotiana benthamiana Infected with tobamoviruses," Proteomics, Vol. 4, No. 20, 2004, pp. 418-425.

110. G. L. Coaker, B. Willard, M. Kinter, E. J. Stockinger and D. M. Francis, "Proteomic Analysis of Resistance Mediated by Rcm 2.0 and Rcm 5.1, Two Loci Controlling Resistance to Bacterial Canker of Tomato," Molecular Plant–Microbe Interactions, Vol. 17, No. 9, 2004, pp. 1019-1028.
111. P. M. Houterman, D. Speijer, H. L. Dekker, C. G. De Koster, B. J. C. Cornelissen and M. Rep, "The Mixed Xylem Sap Proteome of Fusarium oxysporum-infected Tomato Plants," Molecular Plant Pathology, Vol. 8, No. 2, 2007, pp. 215-221.
112. M. A. Castillejo, A. M. Maldonado, E. Dumas-Gaudot, M. Fernández-Aparicio, R. Susin, R. Diego and J. V. Jorrin, "Differential Protein Expression to Investigate Responses and Resistance to Orobranche crenata in Medicago truncatula," BMC Genomics, Vol. 10, 2009, pp. 294-298.
113. F. Delalande, C. Carapito, J. P. Brizard, C. Brugidou and A. Van Dorsselaer, "Multigenic Families and Proteomics: Extended Protein Characterization as a Tool for Paralog Gene Identification," Proteomics, Vol. 5, No. 2, 2005, pp. 450-460.
114. M. Ventelon-Debout, F. Delalande, J. P. Brizard, H. Diemer, A. van Dorsselaer and C. Brugidou, "Proteome Analysis of Cultivar-Specific Deregulations of Oryza sativa indica and O. sativa japonica Cellular Suspension Undergoing rice yellow mottle virus infection," Proteomics, Vol. 4, No. 1, 2004, pp. 216-225.
115. Z. Wei, W. Hu, Q. lin, X. Cheng, M. Tong, L. Zhu, R. Chen and G. He, "Understanding Rice Plant Resistance to the Brown Planthopper (Nilaparvata lugens): A Proteomic Approach," Proteomics, Vol. 9, No. 10, 2009, pp. 2798-2808.
116. S. T. Kim, S. G. Kim, D. H. Hwang, S. Y. Kang, H. J. Kim, B. H. Lee, J. J. Lee and K. Y. Kang, "Proteomic Analysis of Pathogen-Responsive Proteins from Rice Leaves Induced by Rice Blast Fungus, Magnaporthe grisea," Proteomics, Vol. 4, No. 11, 2004, pp. 3569-3578.
117. T. Mahmood, A. Jan, M. Kakishima and S. Komatsu, "Proteomic Analysis of Bacterial-Blight Defence Responsive Proteins in Rice Leaf Blades," Proteomics, Vol. 6, No. 22, 2006, pp. 6053-6065.
118. A. Campos, G. da Costa, A. V. Coelho and P. Fevereiro, "Identification of Bacterial Protein Markers and Enolase as a Plant Response Protein in the Infection of Olea europaea subsp. Pseudomonas savastanoi pv.

Savastanoi," European Journal of Plant Pathology, Vol. 125, No. 4, 2009, pp. 603-616.

119. S. T. Tahara, A. Mehta and Y. B. Rosato, "Proteins Induced by Xanthomonas axonopodis pv. passiflorae with leaf extract of the host plant (Passiflorae edulis)," Proteomics, Vol. 3, No. 1, 2003, pp. 95-102.

120. P. Diaz-Vivancos, M. Rubio, V. Mesonero, P. M. Periago, A. R. Barcelo, P. Martinez-Gomez and J. A. Hernandez, "The Apoplastic Antioxidant System in Prunus: Response to Long-Term Plum Pox Virus Infection," Journal of Experimental Botany, Vol. 57, No. 14, 2006, pp. 3813- 3824.

121. D. Dahal, D. Heintz, A. Van Dorsselaer, H. P. Braun and K. Wydra, "Pathogenesis and Stress Related, as well as Metabolic Proteins Are Regulated in Tomato Stems Infected with Ralstonia solanacearum," Plant Physiology and Biochemistry, Vol. 47, No. 9, 2009, pp. 838-846.

122. L. Babujee, B. Venkatesh, A. Yamazaki and S. Tsuyumu, "Proteomic Analysis of the Carbonate Insoluble Outer Membrane Fraction of the Soft-Rot Pathogen Dickeya dadantii (syn. Erwinia chrysanthemi) Strain 3937," Journal of Proteome Research, Vol. 6, No. 1, 2007, pp. 62-69.

123. C. Rampitsch, N. V. Bykova, B. Mccallum, E. Beimcik and W. Ens, "Analysis of the Wheat and Puccinia triticina (leaf rust) Proteomes during a Susceptible Host Pathogen Interaction," Proteomics, Vol. 6, No. 6, 2006, pp. 1897-1907.

124. W. Zhou, F. Eudes and A. Laroche, "Identification of Differentially Regulated Proteins in Response to a Compatible Interaction between the Pathogen Fusarium graminearum and Its Host, Triticum aestivum," Proteomics, Vol. 6, No. 16, 2006, pp. 4599-4609.

125. S. Campo, M. Carrascal, M. Coca, J. Abian and B. S. Segundo, "The Defence Response of Germinating Maize Embryos against Fungal Infection: A Proteomics Approach," Proteomics, Vol. 4, No. 2, 2004, pp. 383-396.

126. S. Kundu, et al., "Proteomic Analysis of Salicylic Acid Induced Resistance to Mungbean Yellow Mosaic India Virus in Vigna mungo," Journal of Proteomics, Vol. 74, No. 3, 2011, pp. 337-349.

127. P. W. Muturi, J. K. Mwololo, S. W. Munyiri, P. Rubaihayo, J. K. Munyua, M. Mgonja, E. Manyasa and N. Kiarie, "A Perspective on Proteomics: Current Applications, Challenges and Potential Uses," Agriculture and Biology Journal of North America, Vol. 1, No. 5, 2010, pp. 916-918.

128. J. M. Dunwell, A. M. Maria and H. Raul, "Transcriptome Analysis and Crop Improvement," Biological Research, Vol. 34, No. 3-4, 2001, pp. 3-4.

2

CHAPTER

CRYSTALLISATION OF WILD-TYPE AND VARIANT FORMS OF A RECOMBINANT PLANT ENZYME *Β*-*D*-GLUCAN GLUCOHYDROLASE FROM BARLEY (HORDEUM VULGARE L.) AND PRELIMINARY *X*-RAY ANALYSIS

Sukanya Luang [1], James R. Ketudat Cairns [1], Victor A. Streltsov [2] and Maria Hrmova [3],

[1]School of Biochemistry, Institute of Science, Suranaree University of Technology, Nakhon Ratchasima 30000, Thailand

[2]Molecular and Health Technologies, CSIRO-Commonwealth Scientific Research Organization, Victoria 3052, Australia

[3]Australian Centre for Plant Functional Genomics, University of Adelaide, Waite Campus, Glen Osmond, South Australia 5064, Australia

ABSTRACT

Wild-type and variant crystals of a recombinant enzyme β-d-glucan glucohydrolase from barley (Hordeum vulgare L.) were obtained by macroseeding and cross-seeding with microcrystals obtained from native plant protein. Crystals grew to dimensions of up to 500 × 250 × 375 μm at 277 K in the hanging-drops by vapour-diffusion. Further, the conditions are described that yielded the wild-type crystals with dimensions of 80 × 40 ×

60 μm by self-nucleation vapour-diffusion in sitting-drops at 281 K. The wild-type and recombinant crystals prepared by seeding techniques achived full size within 5–14 days, while the wild-type crystals grown by self-nucleation appeared after 30 days and reached their maximum size after another two months. Both the wild-type and recombinant variant crystals, the latter altered in the key catalytic and substrate-binding residues Glu220, Trp434 and Arg158/Glu161 belonged to the $P4_32_12$ tetragonal space group, i.e., the space group of the native microcrystals was retained in the newly grown recombinant crystals. The crystals diffracted beyond 1.57–1.95 Å and the cell dimensions were between a = b = 99.2–100.8 Å and c = 183.2–183.6 Å. With one molecule in the asymmetric unit, the calculated Matthews coefficients were between 3.4–3.5 $Å^3 \cdot Da^{-1}$ and the solvent contents varied between 63.4% and 64.5%. The macroseeding and cross-seeding techniques are advantageous, where a limited amount of variant proteins precludes screening of crystallisation conditions, or where variant proteins could not be crystallized.

Keywords: macro- and cross-seeding; wild-type and mutant protein; X-ray diffraction

1. INTRODUCTION

Plant β-d-glucan glucohydrolase enzymes are classified in the GH3 family of glycoside hydrolases (http://www.cazy.org/) that currently includes nearly 3,000 entries [1]. The majority of these entries represent nucleotide sequences obtained from sequencing of bacterial genomes. Almost all GH3 entries have two, three or more individually folded domains, while the spatial arrangement of the domains varies [2,3]. A large proportion of the GH3 entries are enzymes that are variously annotated as β-d-glucosidases, glucan-1,4-β-d-glucosidases, (1,3)-β-d-glucan exohydrolases, (1,3;1,4)-β-d-glucan exohydrolases, exo-1,3-1,4-β-d-glucanases, N-acetyl-β-d-glucosaminidases, xylan-1,4-β-d-xylosidases and α-l-arabinofuranosidases [1]. Although the GH3 family contains predominantly enzymes, their substrate specificities have not been satisfactorily defined and the most of annotations are based on sequence similarities with enzymes for which biochemical data are available [4].

We have focused our attention in the past upon the biochemical and structural characterization of β-d-glucan glucohydrolases from higher plants, in particular on barley β-d-glucan glucohydrolase, isoenzyme ExoI (designated as HvExoI). This enzyme was crystallized from a native plant source obtained from barley seedlings [5], although the procedure for isolation of a crystallisable quality of HvExoI is both time consuming and technically challenging [6]. Over the past ten years several structures (Protein Data Bank references 1EX1, 1IEQ, 1IEV, 1IEW, 1IEX, 1J8V, 1LQ2, 1X38, 1X39) have been solved from the native HvExoI crystals with substrate analogues and mechanism-based inhibitors to explain the enzyme's catalytic mechanism and substrate specificity [2,7–10]. However, a few facets of catalytic and substrate binding mechanisms remain to be explained. To this end we are interested in the specific roles of amino acid residues in the vicinity of the catalytic pair and how specific structural determinants at the entrance of the catalytic pocket control spatial dispositions of isomeric oligosaccharides entering the catalytic site. Further, we have been intrigued by so-called substrate-product trafficking events at the mouth of the catalytic site [2,7]. Here, the glucose product of the hydrolytic reaction, which is released from the non-reducing termini of substrates, remains bound to the enzyme's active site. However, when a new substrate molecule approaches the enzyme, the glucose product diffuses away from the pocket and the incoming substrate enters the active site [4]. Both events are linked and a precise mechanism, how they proceed, is not known. We presume that the incoming substrate binds in the vicinity of the active site and could, by some delicate mechanism, instigate a product for new substrate interchange. We believe that this substrate/product interchange event represents an ideal model system to study, how products and incoming substrates interact in and/or near enzymes' catalytic sites in general.

Recently, we have reported the high level expression of recombinant HvExoI (designated as rHvExoI) from a codon-optimized HvExoI cDNA in protease-deficient Pichia pastoris under low temperature conditions [11]. We expect that rHvExoI could be used as a suitable enzyme model to study the roles of amino acid residues in catalysis and substrate specificity of this enzyme. To this end, we prepared variants of wild-type rHvExoI, altered (into Ala) in the key catalytic and substrate binding residues Glu220, Trp434 and Arg158/Glu161 by site-directed mutagenesis [12]. From the available structural data [2,7–10] we would expect that Glu220, Trp434 and Arg158 are surface exposed, while Glu161 is buried.

In the current work, we describe the techniques and conditions for preparation of the wild-type and variant rHvExoI crystals that showed excellent diffraction parameters. These crystals were prepared by the macroseeding and cross-seeding techniques with microcrystals obtained from a native plant protein. Seeding techniques have previously been used successfully in bio-macromolecular crystallography [13]. The advantage of these techniques is that they use crystals that provide a preformed, regular surface, onto which new molecules may be added in a regular fashion, generally at a lower degree of supersaturation than is required for nucelation [13].

2. RESULTS AND DISCUSSION

2.1. Protein Expression and Purification

Protein expression in P. pastoris at a temperature of 293 K, deglycosylation and purification yielded near-homogenous recombinant rHvExoI (Figure 1) with a molecular mass of 67,169 Da [11]. It is of note that rHvExoI contained at its NH_2-terminus an 8x-His-tag and additional three Ala residues flanking the 8x-His-tag (AHHHHHHHHAA). The Ala residues resulted from the ligation-based cloning process, while the 8x-His-tag was added for protein purification purposes [11]. This 11-residue tag was not removed from the rHvExoI protein that was subjected to crystallisation trials. Expressed rHvExoI was catalytically competent with the catalytic efficiency value $k_{cat} \cdot K_M^{-1}$ = 14 $mM^{-1} \cdot s^{-1}$ towards 4-nitrophenyl β-d-glucoside. This second-order rate constant was similar to that reported for native fully N-glycosylated HvExoI [14] and a recombinant N-deglycosylated form [11]. Further, the biophysical properties of rHvExoI such as pH optimum and thermostability were also similar to those reported for native or recombinant HvExoI [11,14].

2.2. Protein Crystallisation

The concentrated rHvExoI was subjected to four types of crystallisation trials (Figures 2 and 3). We firstly set-up the microbatch under paraffin oil trials at 277 K and 287 K with the goal to screen approximately 130 independent conditions, including those that were previously found to be successful with a native plant enzyme [5]. These conditions (1.7 M ammonium sulfate, 75 mM HEPES-NaOH buffer, pH 7, containing 7.5 mM sodium acetate and 1.2% (w/v) PEG 400) previously produced crystals of native plant HvExoI that belonged to the primitive tetragonal space group $P4_32_12$ and yielded high-resolution diffraction patterns [2,5,7]. Nevertheless, despite intensive effort, no diffraction-quality crystals grew, and after approximately 14 days only the formation of a highly intricate interlaced web of thin needles of approximately 40–60 μm in their longest dimensions was observed. Figure 2A and its inset show the appearance of these needles that grew from 1.6 M magnesium sulfate at pH 6.5 and 277 K.

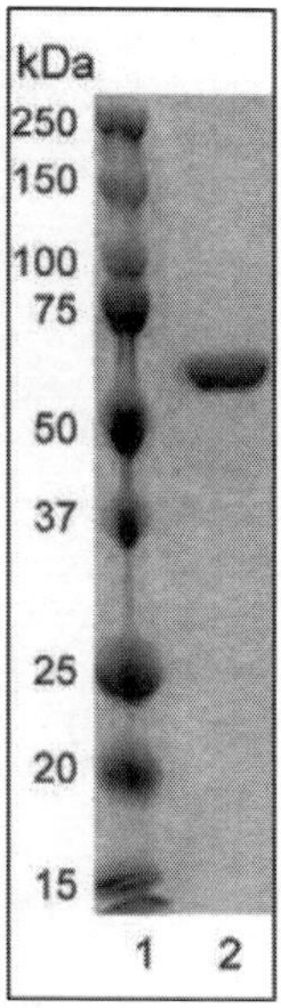

Figure 1. SDS-PAGE profile of recombinant wild-type rHvExoI (lane 2) used for crystallisation by self-nucleation and macro- and cross-seeding using hanging-drop and sitting-drop vapour-diffusion methods. The rHvExoI protein (25 μg) was visualised with a Coomassie Brilliant Blue stain. Protein standards (5 μL of the Precision Plus Protein Standards' stock) are indicated in lane 1.

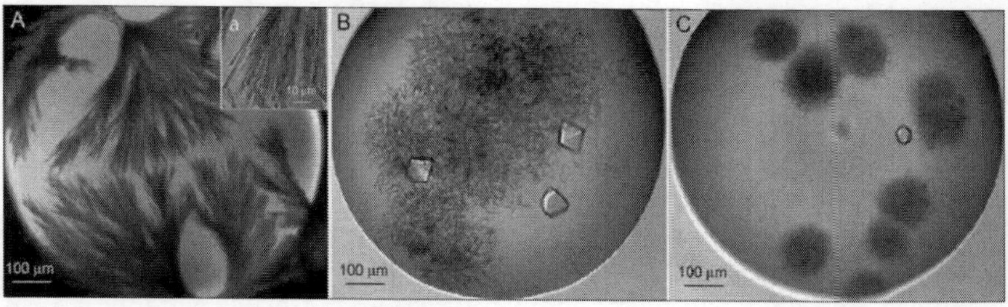

Figure 2. Crystals of recombinant wild-type rHvExoI grown by self-nucleation. Various forms of crystals grown for 14 (**A**) and 97 (**B–C**) days using a microbatch-under-paraffin-oil technique from 1.6 M magnesium sulfate, pH 6.5 at 277 K (A) or obtained using sitting-drop vapour-diffusion from 1.8 (B) or 2.2 M (C) ammonium sulfate, both at pH 5 and 281 K. The crystals in (A) formed thin needles of approximately 40–50 μm in their longest dimensions (inset), while in B and C short thin needles are shown that appeared within 14 days and after another 30 days the truncated bi-pyramidal crystals grew that reached dimensions of 80 × 40 × 60 μm after 97 days.

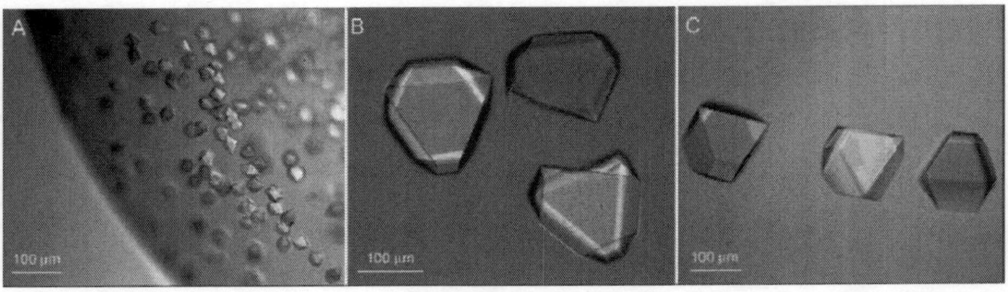

Figure 3. Microcrystals (10 × 5 × 7.5 μm to 20 × 10 × 15 μm) of native HvExoI (**A**) used to grow the recombinant wild-type (**B**) and variant Glu220Ala (**C**) rHvExoI crystals that grew to their full-sizes within 5–14 days. The crystals obtained by seeding in hanging-drops by vapour-diffusion grew to dimensions of up to 500 × 250 × 375 μm at 277 K.

Our second approach to growing the diffraction quality crystals led us to set-up close to 400 trials at the Bio21 Collaborative Crystallisation Centre at 281 K and 293 K by a sitting-drop vapour-diffusion method. After approximately 21 days at 281 K, we could observe formation of short thin needle-shaped crystals that either remained dispersed throughout the droplets (Figure 2B) or formed well-organised round balls (Figure 2C). However, after about a month, in some of the droplets with needles (Figures 2B and 2C) truncated bi-pyramidal crystals formed, which reached dimensions of 80 × 40 × 60 μm after about 97 days (Figure 2B). It was of note that these crystals were only observed at 281 K in droplets with 1.6 to 2.2 M ammonium sulfate, containing 10 mM malate-MES-Tris buffer, pH 5. These conditions were similar to the conditions that we found previously for native HvExoI [5], except that here the pH value of 5 was more acidic then that used previously (pH 7), and that the protein concentration was almost twice as high (12.5 mg·mL^{-1} versus 6.8 mg·ml^{-1}). We expected that these truncated bi-pyramidal crystals could belong to the tetragonal $P4_32_12$ space group found for the native protein crystals [5], although at this stage we were not able to collect their diffraction patterns. Notably, in an identical screen at a higher temperature of 293 K the bi-pyramidal crystals were not formed, so it seemed that the temperature was a critical factor for crystal formation of rHvExoI. Thirdly, we also attempted to grow crystals by self-nucleation at 277 K by hanging-drop vapour diffusion and using the conditions developed for native HvExoI [5]. However, we could not observe crystal formation within 180 days and rHvExoI mostly precipitated or the drops remained clear.

As no diffraction-quality crystals were obtained with the three crystallisation approaches described above, we turned our attention to seeding at 277 K using a hanging-drop vapour-diffusion method that was used successfully for preparation of large well-diffracting crystals of native HvExoI [7–10]. We first examined, if we could use microcrystals obtained from native HvExoI to seed the wild-type recombinant rHvExoI protein, despite the differences between the two proteins. These differences included the 11 additional residues of the affinity tag at the NH_2-terminus, as described above. Further, the rHvExoI protein was N-deglycosylated by endoglycosidase H, such that only one N-linked N-acetyl-β-d-glucosaminyl residue [12] remained attached to each of the three N-glycosylation sites Asn221, Asn498 and Asn600 [2,11]. Approximately 48 h after rHvExoI was macroseeded with the native microcrystals of the sizes between 10 × 5 × 7.5 μm and 20 × 10 × 15 μm, the original native microcrystals started growing in size. The fully grown crystals of wild-type rHvExoI reached dimensions that varied between 100 × 50 × 75 μm (Figure 3B) and 500 × 250 × 375 μm after 5 to 7 days and these crystals had a similar bi-pyramidal morphology as the native microcrystals (Figures 3A). Having succeeded in growing wild-type crystals, using native microcrystal seeds, we prepared large recombinant crystals of three rHvExoI variants, specifically Glu220Ala (Figure 3C), Trp434Ala and of the double mutant Arg158Ala/Glu161Ala, also using native seed crystals. The newly grown variant crystals reached sizes of 100 × 50 × 75 μm (Figure 3C) to 500 × 250 × 375 μm and showed a similar bi-pyramidal morphology as the native microcrystals. The variant crystals grew slightly slower and reached their maximum dimensions after 10–14 days.

2.3. X-ray Diffraction

Single wild-type and variant rHvExoI crystals were cryo-protected and subjected to diffraction at the MX1 beamline of the Australian Synchrotron. All of the X-ray diffraction data sets were virtually complete beyond 1.57–1.95 Å (Figure 4; Table 1). The HKL2000 indexing and systematic absences calculated that the space groups of the wild-type and variant rHvExoI crystals were consistent with a primitive tetragonal space group $P4_32_12$ (Figure 4 and Table 1), and as expected, these space group characteristics were similar to their native counterparts [2,7]. In line with these observations was our previous finding that the native HvExoI microcrystals of the sizes 20 × 10 × 15 μm diffracted beyond 2.80 Å on an in-house rotating anode X-ray source and belonged to a primitive tetragonal space group $P4_32_12$ [15]. Thus, our data are in agreement with other reports, where the space group characteristics of macroseeds and fully grown crystals were identical [16–18], or that as a bonus, the resolution of the newly grown crystals has improved [18]. On the contrary, other authors have reported that during seeding trials mutant crystals often crystallized in different space groups than their macroseeds [19].

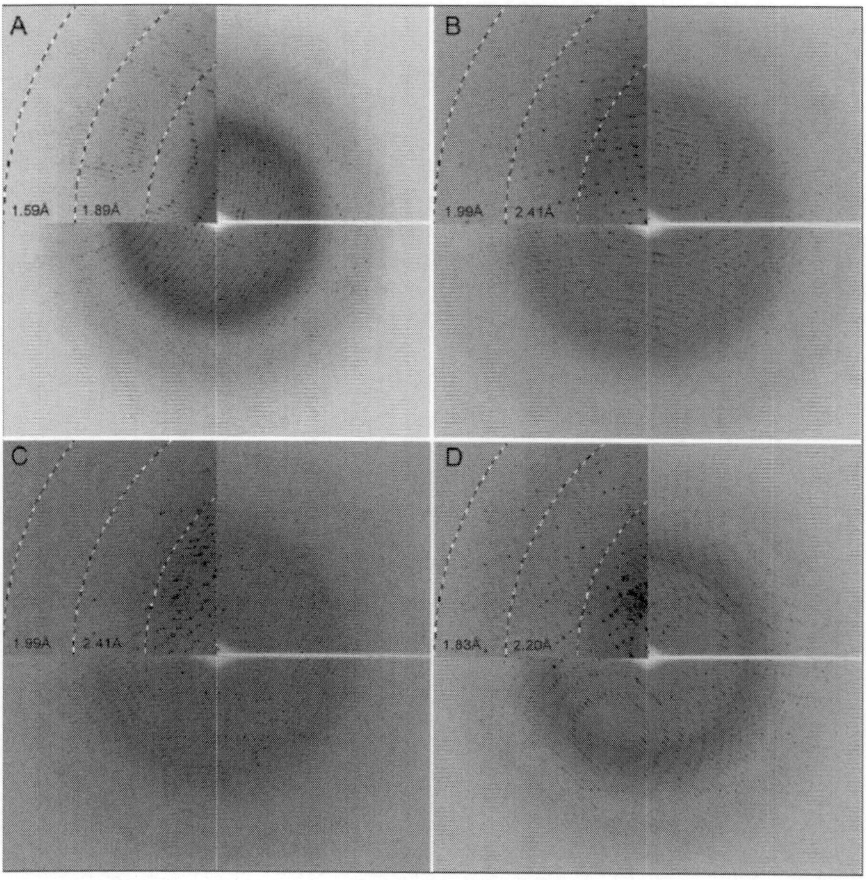

Figure 4. X-ray diffraction patterns of the recombinant wild-type (**A**), and variant Glu220Ala (**B**), Trp434Ala (**C**) and Arg158Ala/Glu161Ala (**D**) rHvExoI crystals. The left top insets show diffraction intensities to 1.59–1.89 Å (A), 1.99–2.41 Å (B and C) and 1.83–2.20 Å (D).

The best diffraction data, to 1.57 Å were collected from the wild-type crystals, followed by the Arg158Ala/Glu161Ala, Glu220Ala and Trp434Ala variants that diffracted to 1.65 Å, 1.90 Å and 1.95 Å, respectively (Figure 4, Table 1). The lattice dimensions of wild-type and variant crystals varied between a = b = 99.2–100.8 Å and c = 183.2–183.6 Å, and there appeared to be one molecule in the asymmetric units, according to the Matthews coefficient calculation [20]. The Matthews coefficients of recombinant crystals were between 3.4–3.5 $Å^3 \cdot Da^{-1}$ with solvent contents of 63.4% to 64.5%. The R_{merge} values of 5.6% to 10.1% were obtained with $<I/\sigma/(I)>$ of

37.0 to 71.2, whereas the multiplicity of the individual datasets was well above 20 and varied between 26 to 29 (Table 1). It was of note that the completeness for the highest resolution shells of the wild-type and Glu220Ala datasets was lower, despite the high multiplicity and crystal symmetry, because the data were integrated into the corners of the detector. Also, the mean <I/σ/(I)> values for these datasets indicated that they actually diffracted to higher resolution than that stated in Table 2. It was not surprising that the most favourable diffraction statistics was obtained with the wild-type crystals that were seeded with the wild-type native macroseeds (Figure 4 and Table 1).

Table 1. Data collection statistics from the wild-type and variant rHvExoI crystals, calculated by the HKL2000 suite of programs.

	Wild-type	Glu220Ala	Trp434Ala	Arg158Ala/Glu161Ala
Unique reflections	119968	74171	64460	107601
Resolution [a] (Å)	1.57 (1.6-1.57)	1.90 (1.94-1.90)	1.95 (1.98-1.95)	1.65 (1.68-1.65)
Mean multiplicity [a,b]	29 (26)	27 (16)	27 (16)	26 (12)
Completeness [a,b] (%)	99 (86)	99 (88)	100 (100)	99.8 (98)
Mean <I/σ/(I)> [a]	71.2 (5.6)	54.8 (4.3)	37.0 (2.2)	58.7 (2.3)
R_{merge} [a-c] (%)	6.7 (47)	8.9 (64)	10.1 (87)	5.6 (82)
a = b (Å) [d]	99.2	100.2	100.1	100.8
c (Å) [d]	183.5	183.2	183.6	183.2

[a] Numbers in parenthesis represent the values in the highest resolution shell.

[b] The numbers were rounded to no decimal place.

[c] $R_{merge} = 100[\sum(I_i-<I>)^2/\sum I_i^2]$ is summed over all independent reflections.

[d] The numbers were rounded to the 1st decimal place.

3. EXPERIMENTAL SECTION

3.1. Expression and Purification of Wild-Type and Variant Forms of rHvExoI

Wild-type (GenBank accession No. GU441535) and variant codon-optimized cDNAs, encoding a mature barley β-d-glucan exohydrolase I (HvExoI) inserted in pPICZαBNH8 expression vectors, were expressed in P. pastoris, strain SMD1168H and purified by ion exchange, immobilized metal affinity chromatography (IMAC), N-deglycosylation by endoglycosidase H and a second round of IMAC, as described previously [11]. The variant forms included Glu220Ala, Trp434Ala and a double mutant Arg158Ala/Glu161Ala, for which the constructions of the cDNA fusions will be described elsewhere [12]. At the final purification step before crystallisation, the N-deglycosylated rHvExoI wild-type and variant enzymes were eluted from a BioGel-P100 size-exclusion column with 50 mM sodium acetate buffer, pH 5.25 containing 200 mM NaCl and 1 mM dithiothreitol at a liner flow rate of 0.5 cm·h^{-1}. The protein purities of the rHvExoI fractions were analyzed by SDS-PAGE, using 12.5% w/v polyacrylamide and bis-polyacrylamide gels and standard techniques [6]. The protein concentration was estimated with a Bio-Rad protein assay kit (Bio-Rad Laboratories, Gladesville, New South Wales, Australia) using bovine serum albumin (Sigma Chemical Company, St. Louis, MO, USA) as a standard. The protein standards 'Precision Plus Protein Standards' used for SDS-PAGE were from Bio-Rad Laboratories.

3.2. Enzyme Assays

The activities of pooled and concentrated (using 10 kDa cut-off centrifugal filter units (Millipore, Bedford, MA, USA) wild-type and variant enzymes were assayed against 4-nitrophenyl β-d-glucopyranoside (Sigma) in 50 mM sodium acetate buffer, pH 5.25.

3.3. Crystallisation of Wild-Type and Variant Forms of rHvExoI

Near-homogenous N-deglycosylated wild-type and variant rHvExoI proteins were concentrated to 12.5 mg·mL^{-1} in 20 mM sodium acetate pH 5.25 and filtered through a 0.22 μm filter (Millipore). Screening of crystallisation conditions was performed by four experimental approaches. Firstly, initial crystallisation conditions were screened using a microbatch under paraffin oil technique. Here, 1 μL of precipitant solutions and 1 μL of the solution containing 12.5 mg·mL^{-1} of rHvExoI were added into 10 μL of 100% Paraffin oil (Hampton Research, Aliso Viejo, CA, USA) that was previously dispensed in microbatch 72 well Greiner (Terasaki style) plates (Hampton Research). The formulations of Crystal Screen 2, Crystal Screen Lite and Grid Screen Ammonium Sulfate (Hampton Research) were used as precipitants and the crystals grew at 277 K or 287 K in a vibration-free crystallographic cabinet (Molecular Dimensions, Suffolk, UK). Secondly, crystallisation trials were set-up robotically (Phoenix Nano-Dispenser, Art Robbins Instruments, Sunnyvale, CA, USA) in sitting drops, in which 300 nL droplets of rHvExoI were mixed with the same volumes of precipitants and crystal growth proceeded at 281 K and 293 K at the Bio21 Collaborative Crystallisation Centre (CSIRO, Parkville, Australia) [21]. The precipitants from the PSS_1_Com5 and PS gradient-mid range formulation screens were used, whereas both screens were prepared in-house at the Bio21 Centre (http://www.csiro.au/c3/Facility/c3_centre_robotic_crystal.htm), following the recommendation from Emerald BioSystems (Bainbridge Island, WA, USA) for the first screen and those reported by Newman [22] for the second screen. The PSS_1_Com5 screen uses the inorganic precipitants such as sulfates, chlorides, citrates and phosphates, but also 2-methyl-2,4-pentanediol, glycerol and polyethyleneglycols (PEGs) in the pH ranges of 5–8.7. On the other hand, the PS gradient-mid range formulation relies on ammonium sulfate and sodium malonate as precipitants in a 10 mM malate-MES-Tris buffer system in the pH ranges of 4.5–9 [22]. Thirdly, attempts were made to grow crystals under the conditions developed for native HvExoI at 277 K using a hanging-drop vapour diffusion method [5]. Lastly, and most importantly, the rHvExoI crystals were grown in hanging drops at 277 K that were seeded with the native HvExoI microcrystals prepared as described previously [5]. The sizes of the microcrystals for the latter conditions varied between 10 × 5 × 7.5 μm and 20 × 10 × 15 μm. The hanging drops were prepared at 277 K as follows. The volume of 4 to 6 μL of rHvExoI at 12.5 mg·mL^{-1} was added to 4 μL of the precipitant solution A (100 mM HEPES-NaOH buffer pH 7, 2.4% (w/v) PEG 400, 1.6 M ammonium sulfate) on 22-mm siliconized circular glass cover slips (Hampton Research). A few microcrystals of native HvExoI were transferred into the hanging drop with a cat whisker. Here, the whisker gently touched the surface of a macroseed stock of native HvExoI and subsequently the whisker was swiftly immersed into a new rHvExoI drop. The cover slips with the seeded hanging drops were placed over 1 mL of reservoir solutions (1.7 M ammonium sulfate in 50 mM HEPES-NaOH buffer, pH 7) contained in the 24 well Linbro plates (Hampton Research), and the wells were sealed with vacuum grease (Dow Corning Corporation, Midland, MI, USA). Crystals from the seeded drops appeared within 5–14 days and were suitable for X-ray data collection. The crystals were photographed through a Leica Laser Microdissection microscope (Leica, North Ryde, Australia) equipped with fluorescence and differential interference contrast.

3.4. X-ray Data Collection and Processing

Single enzyme crystals with the longest dimensions of 100 to 500 μm were cryo-protected in 20% (v/v) glycerol concentration in solution A (as specified above in Section 3.3.) [10] and flash cooled in the cold N_2 stream at the beamline MX1 of the Australian Synchrotron. X-ray diffraction data sets were collected at 0.5° oscillations (1 sec exposures) through 360° on the ADSC Quantum 210r Detector [23]. The data were processed with the HKL2000 suite of programs [24].

4. CONCLUSIONS

In summary, excellent X-ray diffraction data were obtained from the recombinant wild-type and variant rHvExoI crystals grown by seeding from a native plant source protein in hanging-drops by vapour-diffusion. The recombinant crystals grew relatively fast and within 5–14 days reached dimensions of up to 500 × 250 × 375 μm. The fully grown recombinant crystals retained the space group characteristics of their native macroseeds and diffracted beyond 1.57 Å to 1.95 Å. As reported for other proteins, this technique could be valuable, where a limited amount of variant proteins is available precluding crystallisation trials, or where variant protein forms could not be crystallized. We project that cross-seeding using native protein as a source of microcrystals could be successfully used for generation of large recombinant wild-type and variant crystals that could potentially yield high resolution diffraction patterns. Lastly, the diffraction data collected from the wild-type and variant rHvExoI crystals reported here could be used successfully for structure solution. The structural data are currently being prepared for publication [12].

ACKNOWLEDGMENTS

The authors acknowledge the MX1 beamline at the Australian Synchrotron (Victoria, Australia) and Janet Newman from the Bio21 Collaborative Crystallisation Centre (Melbourne, Australia). We are also grateful to Gwenda Mayo (Australian Centre for Plant Functional Genomics) for taking photographs of crystals. This work was supported by grants from the Australian Research Council to M.H., the Thailand Research Fund (TRF) to J.R.K.C. and a TRF Royal Golden Jubilee PhD Fellowship to S.L. The constructive help of both anonymous referees is also acknowledged.

REFERENCES

1. Cantarel, BL; Coutinho, PM; Rancurel, C; Bernard, T; Lombard, V; Henrissat, B. The Carbohydrate-Active EnZymes database (CAZy): An expert resource for glycogenomics. Nucleic Acids Res **2008**, 37, D233–D238.
2. Varghese, JN; Hrmova, M; Fincher, GB. Three-dimensional structure of a barley β-d-glucan exohydrolase; a family 3 glycosyl hydrolase. Struct. Fold. Des **1999**, 7, 179–190.
3. Pozzo, T; Pasten, JL; Karlsson, EN; Logan, DT. Structural and functional analyses of beta-glucosidase 3B from Thermotoga neapolitana: A thermostable three-domain representative of glycoside hydrolase 3. J. Mol. Biol **2010**, 397, 724–739.
4. Hrmova, M; Fincher, GB. Dissecting the catalytic mechanism of a plant β-d-glucan glucohydrolase through structural biology using inhibitors and substrate analogues. Carbohydr. Res **2007**, 305, 209–221.
5. Hrmova, M; Varghese, JN; Høj, PB; Fincher, GB. Crystallization and preliminary X-ray analysis of β-glucan exohydrolase isoenzyme ExoI from barley (Hordeum vulgare). Acta Cryst **1998**, D54, 6887–689.

6. Hrmova, M; Harvey, AJ; Wang, J; Shirley, NJ; Jones, GP; Høj, PB; Fincher, GB. Barley β-d-glucan exohydrolases with β-d-glucosidase activity. Purification and determination of primary structure from a cDNA clone. J. Biol. Chem **1996**, 271, 5277–5286.
7. Hrmova, M; Varghese, JN; DeGori, R; Smith, BJ; Driguez, H; Fincher, GB. Catalytic mechanisms and reaction intermediates along the hydrolytic pathway of plant β-d-glucan glucohydrolase. Struct. Fold Des **2001**, 9, 1005–1016.
8. Hrmova, M; DeGori, R; Smith, BJ; Driguez, H; Varghese, JN; Fincher, GB. Structural basis for a broad specificity in higher plant β-d-glucan glucohydrolases. Plant Cell **2002**, 14, 1033–1052.
9. Hrmova, M; De Gori, R; Smith, BJ; Vasella, A; Varghese, JN; Fincher, GB. Three-dimensional structure of the barley β-d-glucan glucohydrolase in complex with a transition state mimic. J. Biol. Chem **2004**, 279, 4970–4980.
10. Hrmova, M; Streltsov, VA; Smith, BJ; Vasella, A; Varghese, JN; Fincher, GB. Structural rationale for low nanomolar binding of transition state mimics to a family GH3 β-d-glucan glucohydrolase from barley. Biochemistry **2005**, 44, 16529–16539.
11. Luang, S; Hrmova, M; Ketudat Cairns, JR. High-level expression of barley β-d-glucan exohydrolase HvExo I from a codon-optimized cDNA in Pichia pastoris. Prot. Expr. Purif **2010**, 73, 90–98.
12. Luang, S; Streltsov, VA; Ketudat Cairns, JR; Hrmova, M. Characterization of variant forms of β-d-glucan glucohydrolase from barley (Hordeum vulgare L). Biochem J, 2010. to be submitted..
13. Stura, E; Wilson, IA. Analytical and production seeding techniques. Methods **2000**, 1, 38–49.
14. Hrmova, M; Fincher, GB. Barley β-d-glucan exohydrolases. Substrate specificity and kinetic properties. Carbohydr. Res **1998**, 305, 209–221.
15. Varghese, JN; van Donkelaar, A; Balaic, DX; Barnea, Z. CSIRO-Commonwealth Scientific Research Organization. Personal communication: Melbourne, Victoria, 2001.
16. Radaev, S; Agniswamy, J; Sun, PD. A case of structure determination using pseudosymmetry. Acta Cryst **2009**, D65, 1334–1340.
17. Stec, B; Holtz, KM; Wojciechowski, CL; Kantrowitz, ER. Structure of the wild-type TEM-1 β-lactamase at 1.55 Å and the mutant enzyme Ser70Ala at 2.1 Å suggest the mode of noncovalent catalysis for the mutant enzyme. Acta Cryst **2005**, D61, 1072–1079.

18. Heaslet, H; Rosenfeld, R; Giffin, M; Lin, Y-C; Tam, K; Torbett, BE; Elder, JH; McRee, DE; Stout, CD. Conformational flexibility in the flap domains of ligand free HIV protease. Acta Cryst **2007**, D63, 866–875.
19. Mizutani, H; Saraboji, K; Malathy Sony, SM; Ponnuswamy, MN; Kumarevel, T; Krishna Swamy, BS; Simanshu, DK; Murthy, MR; Kunishima, N. Systematic study on crystal-contact engineering of diphthine synthase: Influence of mutations at crystal-packing regions on X-ray diffraction quality. Acta Cryst **2008**, D64, 1020–1033.
20. Matthews, BW. Solvent content of protein crystals. J. Mol. Biol **1968**, 33, 491–497.
21. Newman, J; Pham, TM; Peat, TS. Phoenito experiments: combining the strengths of commercial crystallization automation. Acta Cryst **2008**, F64, 991–996.
22. Newman, J. Optimization of buffer solutions for protein crystallization. Acta Cryst **2004**, D60, 610–612. [Google Scholar]
23. McPhillips, TM; McPhillips, SE; Chiu, HJ; Cohen, AE; Deacon, AM; Ellis, PJ; Garman, E; Gonzalez; Sauter, NK; Phizackerley, RP; Soltis, SM; Kuhn, P. Blu-ice and the distributed control system: Software for data acquisition and instrument control at macromolecular crystallography beamlines. J. Synchrotron Radiat **2002**, 9, 401–406.
24. Otwinowski, Z; Minor, W. Processing of X-ray diffraction data collected in oscillation mode. Methods Enzymol **1997**, 276, 307–326.

3

CHAPTER

EFFICACY OF PLANT EXTRACTS IN PLANT DISEASE MANAGEMENT

Malkhan Singh Gurjar[1], Shahid Ali[2*], Masood Akhtar[2], Kangabam Suraj Singh[1]

[1]*Central Potato Research Station, Shillong, India*

[2]*SNPG College, Azamgarh (VBS Purvanchal University, Jaunpur), India;*

ABSTRACT

The overzealous and indiscriminate use of most of the synthetic fungicides has created different types of environmental and toxicological problems. Recently, in different parts of the world, attention has been paid towards exploitation of higher plant products as novel chemotherapeutants in plant protection. The popularity of botanical pesticides is once again increasing and some plant products are being used globally as green pesticides. Pyrethroids and neem products are well established commercially as botanical pesticides and recently some essential oils of higher plants have also been used as antimicrobials against storage pests because of their relatively safe status and wide acceptance by the consumers. Some of the volatile oils, which often contain the principal aromatic and flavouring components of herbs and spices, have been recommended as plant based antimicrobials to retard microbial contamination and reduction in spoilage of food commodities. In the context of agricultural pest management, botanical pesticides are best suited for use in organic food production in industrialized countries but can play a much greater role in the production and post harvest protection of food products in developing countries.

Keywords: Botanical Pesticides; Plant Protection; Chemotherapeutants; Organoleptic

1. INTRODUCTION

The ultimate aim of recent research in this area has been the development of alternative control strategies to reduce dependency on synthetic fungicides. Plants have ability to synthesize aromatic secondary metabolites, like phenols, phenolic acids, quinones, flavones, flavonoids, flavonols, tannins and coumarins [1]. The components with phenolic structures, like carvacrol, eugenol, and thymol, were highly active against the pathogen. These groups of compounds show antimicrobial effect and serves as plant defence mechanisms against pathogenic microorganisms [2]. The volatile antimicrobial substance allicin (diallyl thio sulphinate) is synthesized in garlic when the tissues are damaged and the substrate alliin (S-allyl-L-cysteine Sulphoxide) mixes with the enzyme alliin-lyase. Allicin is readily membrane-permeable and undergoes thiol-disulphide exchange reactions with free thiol groups in proteins. Allicin effectively controlled seed-borne Alternaria spp. in carrot, Phytophthora leaf blight of tomato and tuber blight of potato as well as Magnaporthe on rice and downy mildew of Arabidopsis thaliana [3]. Application of plant products especially essential oils is a very attractive method for controlling post harvest diseases. Essential oil extracted from lemon grass (Cymbopogon spp.) post harvest anthracnose of mango fruit [4]. The anti viral protein (AVP) extracts from Bougainvillea spectabilis and Prosopis chilinesis were found to be effective in reducing the sunflower necrosis virus (SFNV) infection both in cowpea and sunflower plants [5]. At present, scientists are investigating for plant products of antimicrobial properties. It would be advantageous to standardize methods of extraction and in vitro antimicrobial efficacy testing so that the search for new biologically active plant products could be more systematic. Thousands of phyto chemicals which have inhibitory effects on all types of microorganisms in vitro should be subjected in vivo testing to evaluate the efficacy in controlling the incidence of diseases in crops, plants, and humans [2].

2. BOTANICALS

Some plant contains components that are toxic to pathogens. When extracted from the plant and applied on infested crops, these components are called botanical pesticides or botanicals.

Commonly used botanicals:

Plant extracts: Neem (Azadirachta indica, A. Juss), Garlic (Allium sativum, Linn., Eucalyptus (Eucalyptus globulus, Labill., Turmeric (Curcuma Longa, Linn., Tobacco (Nicotiana tabacum, Linn., Ginger (Zingiber officinale, Rosc.

Essential oils: Nettle oil (Urtica spp.), Thyme oil (Thymus vulgaris, Linn.), Eucalyptus oil Eucalyptus globulus, Labill. Rue oil (Ruta graveolens, Linn.), Lemon grass oil (Cymbopogon flexuosus (Steud.) Wats. and Tea tree oil (Melaleuca alternifolia).

Gel and latex: Aloe vera (Tourn. Ex Linn.).

- Why Consider Botanicals?
- Sustainable solutions in agriculture
- Reduce crop losses
- Eco-friendly
- Easily bio-degradable
- Organic farming
- Cheaper
- Integrated Diseases Management

3. PLANT EXTRACTS

3.1. Methods of Plant Extract Preparation

Extraction methods involve separation of medicinally active fractions of plant tissue from inactive/inert components by using selective solvents and extraction technology (Table 1). Solvents diffuse into the solid plant tissues and solubilize compounds of similar polarity. Quality of plant extract depends on plant material, choice of solvents and the extraction methods.

3.2. Plant Material

Plants are collected either randomly or by following leads supplied by local healers in geographical areas where the plants are found. Fresh or dried plant materials can be used as a source for the extraction of secondary plant components. Many authors had reported about plant extract preparation from the fresh plant tissues [6]. The logic behind this came from the ethno medicinal use of fresh plant materials among the traditional and tribal people. But as many plants are used in the dry form (or as an aqueous extract) by traditional healers and due to differences in water content within different plant tissues, plants are usually air dried to a constant weight before extraction. Other researchers dry the plants in the oven at about 40°C for 72 h [7]. In most of the reported works, underground parts (root, tuber, rhizome, bulb etc.) of a plant were used extensively compared with other above ground parts in search for bioactive compounds possessing antimicrobial properties.

3.3. Choice of Solvents

Successful determination of biologically active compound from plant material is largely dependent on the type of solvent used in the extraction procedure. Properties of a good solvent in plant extractions include low toxicity, ease of evaporation at low heat, promotion of rapid physiologic absorption of the extract, preservative action and inability to cause the extract to complex or dissociate. As the end product in extraction will contain traces of residual solvent, the solvent should be non-toxic and should not interfere with the bioassay [8]. The choice will also depend on the targeted compounds to be extracted. Initial screening of plants for possible antimicrobial activities typically begins by using the crude or alcohol extractions and can be followed by various organic solvent extraction methods. Water is universal solvent, used to extract plant products with antimicrobial activity. Though traditional healers use primarily water but plant extracts from organic solvents have been found to give more consistent antimicrobial activity compared to water extract [9]. Also water soluble flavonoids (mostly anthocyanins) have no antimicrobial significance and water soluble phenolics only important as antioxidant compound a study reported that extraction of tannins and other phenolics were better in aqueous acetone than in aqueous methanol. In another study, among the twenty different solvents evaluated, chloroform was found to be the best solvent for the extraction of non-polar biological active compounds [10]. Since nearly all of the identified antimicrobial compounds from plants are aromatic or saturated organic compounds, they are most often obtained through initial ethanol or methanol extraction. Thus the most commonly used solvents for preliminary investigations of antimicrobial activity in plants are methanol, ethanol and water. The other solvents used by researchers are dicholro-methane, acetone, and hexane. Some authors use a combination of these solvents to obtain the best solvent systems for extraction. [11] examined a variety of extractants for their ability to solubilize antimicrobials from plants, rate of extraction, ease of removal, toxicity in bioassay and acetone received the highest overall rating. Though there is a wide diversification in the usage of solvents, it is necessary to focus on a standardized extraction method and solvent system for a wide variety of researchers working in diverse settings to minimize the variability in the antimicrobial efficacy reports.

Table 1. Solvents used for active component extraction [1].

Water	Ethanol	Methanol	Chloroform	Dichloro-methanol	Ether	Acetone
Tannins	Alkaloids	Terpenoids	Terpenoids	Terpenoids	Alkaloids	Flavonols
Saponins	Tannins	Saponins	Flavonoids	-	Terpenoids	-
Terpinoides	Terpinoides	Tannins	-	-	Coumarins	-
-	Flavonol	Flavones	-	-	-	-

3.4. Extraction Methods

Variation in extraction methods are usually depend on the length of the extraction period, solvent used, pH of the solvent, temperature, particle size of the plant tissues and the solvent-to-sample ratio. The basic principle is to grind the plant material (dry or wet) finer, which increases the surface area for extraction thereby increasing the rate of extraction. In the study by [12], 5 min extractions of very fine particles of diameter 10 µm gave higher quantities than values obtained after 24 h in a shaking machine with less finely ground material. Earlier studies reported that solvent-to-sample ratio of 10:1 (v/w) solvent to dry weight ratio has been used as ideal [13]. The extraction method that has been widely used by researchers is plant tissue homogenization in solvent [9]. Dried or wet, fresh plant parts are ground in a blender to fine particles, put in a certain quantity of solvent and shaken vigorously for 5 - 10 min or left for 24 h after which the extract is filtered. The filtrate then may be dried under reduced pressure and re-dissolved in the solvent to determine the concentration. Some researchers however centrifuged (approximately 20,000 × g, for 30 min) the filtrate for clarification of the extract. Another common method is serial exhaustive extraction which involves successive extraction with solvents of increasing polarity from a non polar (hexane) to a more polar solvent (methanol) to ensure that a wide polarity range of compound could be extracted [13]. Other researchers employ soxhlet extraction of dried plant material using organic solvent. This method cannot be used for thermolabile compounds as prolonged heating may lead to degradation of compounds.

4. ANTIMICROBIAL SECONDARY METABOLITES

Plants have limitless ability to synthesize aromatic secondary metabolites, most of which are phenols or their oxygen-substituted derivatives.

Important subclasses in this group of compounds include phenols, phenolic acids, quinones, flavones, flavonoids, flavonols, tannins and coumarins (Figure 1). These groups of compounds show antimicrobial effect and serves as plant defense mechanisms against pathogenic microorganisms (Table 2). Simple phenols and phenolic acid are bioactive phytochemicals consisting a single substituted phenolic ring. Phenolic toxicity to microorganisms is due to the site(s) and number of hydroxyl groups present in the phenolic compound. Quinones are characteristically highly reactive, colored compounds with two ketone substitutions in aromatic ring. Flavones, flavonoids and flavonols are phenolic structure with one carbonyl group. They are synthesized by plants in response to microbial infection and are often found effective in vitro as antimicrobial substance against a wide array of microorganisms (Table 3). Tannins are polymeric phenolic substances possessing the astringent property. These compounds are soluble in water, alcohol and acetone and give precipitates with proteins. Coumarins are phenolic substances made of fused benzene and α-pyrone rings. The crude sap, volatile and essential oil extracted from whole plant or specialised plant parts like roots, stem, leaves, flowers, fruits and seeds are widely used in preparing the antimicrobial compounds which are significantly used against the different plant pathogens/diseases (Table 4).

Six broad chemical groups are as follows.

- Flavonoides and isoflavonoides
- Saponins
- Steroides
- Tannins
- Phenolic and phenolic acids—Chlrogenic acid, protocatechuic acid, ferulic acid, caffeic acid
- Coumarins and Pyrones

Table 2. Mode of action of phyto chemicals [1].

Class	Sub-class	Mechanism
Phenolics	Simple phenols	Membrane disruption, substrate deprivation
Phenolic acids	Phenolic acids	Bind to adhesins, complex with cell wall, inactivate enzymes
Terpenoids, essential oils		Membrane disruption
Alkaloids		Intercalate into cell wall
Tannins		Bind to proteins, enzyme inhibition, substrate deprivation
Flavonoids		Bind to adhesins, complex with cell wall, Inactivate enzymes
Coumarins		Interaction with eucaryotic DNA
Lectins and polypeptides		Form disulfide bridges

Table 3. Botanicals produced by plants having antimicrobial activity.

Common name	Scientific name	Compound	Class	Activity
Apple	*Malus pumila* Mill.	Phloretin	Flavonoid derivative	General
Ashwagandha	*Withania somnifera* Dunal.	Withafarin A	Lactone	Bacteria, fungi
Bael tree	*Aegle marmelos* Linn.	Essential oil	Terpenoid	Fungi
Blue gum tree	*Eucalyptus globulus* Labill.	Tannin	Polyphenol	Fungi, Bacteria, Viruses
Onion	*Allium cepa* Linn.	Allicin	Sulfoxide	Fungi, Bacteria
Thyme	*Thymus vulgaris* Linn	Caffeic acid	Terpenoid	Fungi, Bacteria, viruses
Turmeric	*Curcuma longa* Linn.	Curcumin	Terpenoids	Fungi, Bacteria, protozoa
Thorn apple	*Datura stramonium* Linn.	Hyoscymine Scopolamine	Alkaloids	Fungi
Black pepper	*Piper nigrum* Linn.	Piperine	Alkaloid	Fungi
Castorbean	*Ricinus communis* Linn.	Ricinine Ricininoleic	Alkaloids	Fungi
Neem/Margosa tree	*Azadirachta indica* A.Juss.	Azadirachtin	Terpenoides	Fungi, Bacteria
Garlic	*Allium sativum* Linn.	Allicin	Solfoxide	Fungi, Bacteria

Table 4. Botanicals.

Neem/Margosa (*Azadirachta indica* A.Juss.).	Leaf	Achook formulations (azadirachtina)	Sheath blight of rice	[30]
Neem/Margosa (*Azadirachta indica* A.Juss.).	Seed kernel	Neem oil	Rice tungro virus	[31]
Neem/Margosa (*Azadirachta indica* A.Juss.).	Leaf, Seed	Achook, Neemazal,	Bacterial blight of rice	[32]
Oregano (*Origanum hercleoticum* (weed species)	Leaf	Essential oils	*Fusarium oxysporum, Phoma tracheiphila*	[4]
Neem/Margosa (*Azadirachta indica* A.Juss.), Black cumin (*Nigelia sativa* Linn. Asfetida (*Ferula asafoetida* Linn.)	Seeds	Essential oils	*Fusarium oxysporum, A.niger, A.flavus*	[33]
Strawberry (*Fragaria* spp.)	Fruit	Volatile compounds	Anthracnose of strawberry	[34]
Raspberry (*Rubus* spp.) and Strawberry (*Fragaria* spp.)	Fruit	Volatile compounds	Post harvest decay fungi	[35]
Garden croton (*Codiaeum variegatum* Linn.)	Leaf	Phenolic compounds	*Alternaria alternate, Fusarium oxysporum*	[36]
Oleander (Nerium *oleander* Linn.)	Leaf	Crude extracts	Brown spot of rice (*Bipolaris oryzae*)	[37]
Indian aloe (*Aloe barbadensis* Mill.) Neem/Margosa (*Azadirachta indica* A.Juss.), Tobacco (*Nicotiana tabacum* Linn.)	Leaf	Crude extracts	Dry rot of yam *F. oxysporum, A.nizer*	[38]
Black pepper (*Piper nigrum* Linn., Clove (*Syzygium aromaticum* (Linn.) Merr. & Perry, Geranium (*Pelargonium graveolens* L'Herit), Nutmeg (*Myristica fragrans* Houtt.), (*Origanum vulgare* spp. *hirtum* (Link) Letsw. and thyme [*Thymus vulgaris* Linn.	Leaf	Volatile oil	Anti bacterial (gram positive and gram negative)	[39]
Metasequoia glyptostroboides	Leaf	Essential oil	*Fusarium oxysporum, Fusarium solani, Phytophthora capsici, Colletotrichum capsici, Sclerotinia sclerotiorum, Botrytis cinerea* and *Rhizoctonia solani*	[40]
Clove (*Syzigium aromaticum* Linn.), Turmeric (*Curcuma longa* Linn.), garlic (*Allium sativum* Linn.) and Holy Basil (*Ocimum Sanctum* Linn.)	Leaf, seed, fruit	Crude extract	*Aspergillus flavus*	[41]
Aspilia africana, Chromolaena odorata, Musa paradisiaca and *Tithonia diversifolia*	Leaf	Aqueous extract	*Cercospora* leaf spot of Seasame (*Sesamum indicum*) L.)	[42]
Lemon grass (*Cymbopogon* spp.), Thyme (*Thymus vulgaris* Linn.)	Leaf, root	Volatile compound	Black Mould Disease on Onion Bulbs (*Aspergillus niger*)	[43]
Brassica napus and Tomato (*Lycopersicon esculentum* Mill.)	Leaf, stem	Water extract	Bacterial disease on Onions	[44]
Ginger (*Zingiber officinale* Rosc.), aloe (*Aloe* vera), bitter kola (*Garcinia cola*) and neem (*Azadirachta indica* A.Juss.)	Leaf, Fruit and Seed	Crude	Root rot disease of cow pea (*Vigna unguiculata* L.)	[45]
Chloranthus japonicas, Paulownia coreana	Roots, stem	Crude extract	rice blast, rice sheath blight, and wheat leaf rust	[46]
Urtica dioica L., thyme *Thymus vulgaris* L., *Eucalyptus* spp., *Ruta graveolens* L. and *Achillea millefolium* L.	Leaf	Volatile oil	*Aletrnaria alternate*	[47]
Acacia nilotica, Achras zapota, Datura stramonium, Emblica officinalis, Eucalyptus globules, Lawsonia inermis, Mimusops elengi, Peltophorum pterocarpum, Polyalthia longifolia, Prosopis juliflora, Punica granatum and *Sygigium cumini*	Leaf, Seed, fruits	Aqueous extract	*Aspergillus* spp.	[48]
Eugenia aromatica , Piper betle), Alpinia galanga and *Sphaeranthus indicus*	Leaf	Crude	Stem rot disease of Vanilla	[49]
Ocimum gratissimum, Aframomum melegueta	Leaf	Crude	Post harvest yam (*Diascorea* spp.) rot	[50]
Neem (*Azadirachta indica* A.Juss.)	Leaf	Crude	Protozoan, Bacteria, Antifertility	[51]

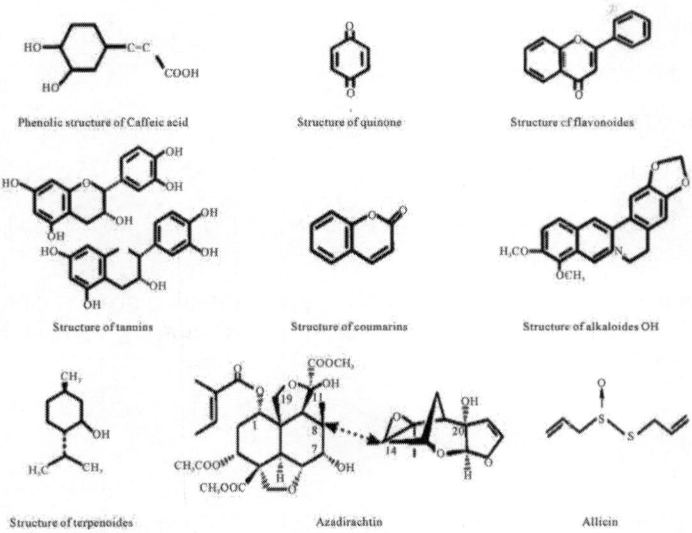

Figure 1. Structure of antimicrobial compounds [14].

Methods for Evaluation of Efficacy of Plant Extract

In vitro antimicrobial susceptibility testing (AST)

1) Diffusion test

- a) agar well diffusion
- b) agar disk diffusion
- c) poison food technique
- d) bio autography

2) Dilution methods

- a) agar dilution
- b) broth micro dilution assay
- c) broth macro dilution assay

5. VOLATILE OILS

- Volatiles are small molecular weight organic compounds having appreciable vapour pressure at ambient temperature

- Play role in defense system
- Inhibit the growth of pathogen
- Generally Recognized As Safe (GRAS)
- Isolated by hydro distillation or steam
- Terpinoides—monoterpenes (C10), ses quiterpenes (C15), diterpines (C20)

Examples: Black pepper (Piper nigrum Linn.), Clove (Syzygium aromaticum Linn.) Merr. & Perry, Nutmeg (Myristica fragrans Houtt.), Oregano (Origanum vulgare Linn.) and Thyme (Thymus vulgaris Linn.)

6. ESSENTIAL OILS

- The potential essential oils used against post harvest pathogens.
- Fungistatic
- Terpenoides and aromatic compounds
- Component—carvacrol, thymol, cymene, terpine, phenylpropene derivatives, eucalyptol and anisol Examples: Nettle oil (Urtica spp.), Thyme oil (Thymus vulgais Linn.), Eucalyptus oil (Eucalyptus globules Labill., Rue oil (Ruta graveolens Linn.), Lemon grass oil (Cymbopogon flexuosus (Steud.) Wats.) and Tea tree oil (Melaleuca alternifolia Maiden & Betche).

7. LIMITATIONS OF BOTANICALS FOR PLANT DISEASE MANAGEMENT

- Extraction methods are not standardized
- Rapid degradation
- Most studies are in vitro efficacy
- Need the development of formulations
- Some chemical compounds are harmful to human and plants
- Less effective
- Less availability formulations

8. CONCLUSION

Plants contain thousands of constituents and are valuable sources of new and biologically active molecules possessing antimicrobial property (Table 4). The ethnobotanical study of plant is important for modern day medicine but its usefulness cannot be overemphasized if methods are not standardized to obtain comparable and reproducible results. At present, scientists are investigating for plant products of antimicrobial properties. It would be advantageous to standardize methods of extraction and in vitro antimicrobial efficacy testing so that the search for new biologically active plant products could be more systematic and interpretation of results would be facilitated. Thousands of phytochemicals which have inhibitory effects on all types of microorganisms in vitro should be subjected in vivo testing to evaluate the efficacy in controlling the incidence of disease in crops, plants, and humans. Efficient collaborations with pharmacologists and medical doctors, plant pathologists and microbiologists are crucial to see the complete development of an interesting lead compound into an exploitable product.

REFERENCES

1. Cowan, M.M. (1999) Plant products as antimicrobial agents. Clinical Microbiology Reviews, 12, 564-582.
2. Das, K., Tiwari, R.K.S. and. Shrivastava, D.K. (2010) Techniques for evaluation of medicinal plant products as antimicrobial agent: Current methods and future trends. Journal of Medicinal Plants Research, 4, 104-111.
3. Slusarenko, A.J., Patel, A. and Portz, D. (2008) Control of plant diseases by natural products: Allicin from garlic as a case study. European Journal of Plant Pathology, 121, 313-322.
4. Salomone, A., Scaritto, G., Sacco, A., Cabras, G. and Angioni, A. (2008) Inhibitory effects of the main compounds of oregano essential oil against some pathogenic fungi. Modern fungicide and Antifungal compounds V. Deutsche Phytomedizinische Gesellschaft, 44, 345-360.
5. Lavanya, N., Saravanakumar, D., Rajendran, L., Ramiah, M., Raguchander, T. and Samiyappan, R. (2009) Management of sunflower necrosis virus through anti-viral substances. Archives of Phytopathology and Plant Protection, 42, 265-276.

6. Tiwari, R.K.S., Chandravanshi, S.S. and Ojha, B.M. (2005) Efficacy of extracts of medicinal plant species on growth of Sclerotium rolfsii root rot in tomato. Journal of Mycology and Plant Pathology, 34, 461-464.
7. Salie, F., Eagles, P.F.K. and Lens, H.M.J. (1996) Preliminary antimicrobial screening of four South African Asteraceae species. Journal of Ethnopharmacology, 52, 27- 33.
8. Ncube, N., Afolayan, S.A.J. and Okoh, A.I. (2008) Assessment techniques of antimicrobial properties of natural compounds of plant origin: Current methods and future trends. African Journal of Biotechnology, 7, 1797-1806.
9. Parekh, J., Jadeja, D. and Chanda, S. (2005) Efficacy of aqueous and methanol extracts of some medicinal plants for potential antibacterial activity. Turkish Journal of Biology, 29, 203-210.
10. Harmala, P., Vuorela, H., Tornquist, K. and Hiltunen, R. (1992) Choice of solvent in the extraction of Angelica archangelica roots with reference to calcium blocking activity. Planta Medica, 58, 176-183.
11. Eloff, J.N. (1998) A sensitive and quick microplate method to determine the minimum inhibitory concentration of plant extracts for bacteria. Planta Medica, 64, 711-713.
12. Eloff, J.N. (1998) Which extractant should be used for the screening and isolation of antimicrobial components from plants? Journal of Ethnopharmacology, 60, 1-8.
13. Green, R.J. (2004) Antioxidant activity of peanut plant tissues. Master Thesis, North Carolina State University.
14. Ciocan, I.D. and Băra, I. (2007) Plant products and antimicrobial agents. Universitatii ale Ştiinţifice Analele Alexandru Ioan Cuza, Tom VIII.
15. Manoharachary, C. and Gourinath, A. (1988) Effects of plant extracts on four pathogenic fungi. 5th International Congress of Plant Pathology, Kyoto.
16. Bandara, B.M.R., Kumar, N.S. and Samaranayake, K.M.S. (1989) An antifungal constituent from the stembark of Butea monosperma. Journal of Ethnopharmacology, 25, 73-75.
17. Noriel, L.M. and Robles, R.P. (1990) Fungicidal activity of Portulaca oleracea extract against Helminthosporium maydis Wisik and Miyake in corn (Zea mays L.). Philippine Journal of Weed Science, 17, 26-32.
18. Natarajan, M.R. and Lalithakumari, D. (1987) Antifungal activity of the leaf extract of Lawsonia inermis on Dreschslera oryzae. Indian Phytopathology, 40, 390-395.

19. Nduagu, C., Ekefan, E.J. and Nwankiti, A.O. (2008) Effect of some crude plant extracts on growth of Colletotrichum capsici (Synd) Butler & Bisby, causal agent of pepper anthracnose. Journal of Applied Biosciences, 6, 184-190.
20. Dharam, V. and Sharma R.K. (1985) Efficacy of fungicides XXIX studies on the fungicidal properties of neem oil. Indian Journal of Plant Pathology, 3, 241-242.
21. Mishra, A.K., Dwivedi, S.K. and Kishore, N. (1989) Antifungal activity of some essential oils. National Academy Science Letters, 12, 335-336.
22. Upadhyaya, M.R. and Gupta, R.C. (1990) Effect of extracts of some medicinal plants on the growth of Curvularia lunata. Indian Journal of Mycology and Plant Pathology, 20, 144-145.
23. Yegen, O., Begger, B. and Heitefuss, R. (1992) Studies on the fungitoxic effect for extracts of six selected plants from Turkey on phytopathogenic fungi. Zeitschrift fur Pflanzenkrankheiten und Pflanzenschütz, 99, 349-359.
24. Bhattiprolu, S.L. and Bhattiprolu, G.R. (2006) Management of castor grey rot disease using botanical and biological agents. Indian Journal of Plant Protection, 34, 101-104.
25. Patil, M.J., Ukey, S.P. and Raut, B.T. (2001) Evaluation of fungicides and botanicals for the management of early blight (Alternaria solani) of tomato. PKV Research Journal, 25, 49-51.
26. Srinivas, T., Rao, M.S., Reddy, P.S. and Reddy, P.N. (1997) Integrated management of leaf spot of groundnut (Arachis hypogaea L.) with botanicals and chemicals. Zeitschrift fur Pflanzenkrankheiten und Pflanzenschutz, 104, 528-530.
27. Surwase, A.G., Badgire, D.R. and Suryawanshi, A.P. (2009) Management of pea powdery mildew by fungicides, botanicals and bio-agents. Annals of Plant Protection Sciences, 17, 384-388.
28. Bhosale, M.V., Perne, R.R., Shete, M.H. and Pawar, N.B. (2008) Efficacy of botanicals, bioagents and fungicides on management of leaf blight (Alternaria porri) of onion. Journal of Plant Disease Sciences, 3, 88-89.
29. Tripathi, P., Dubey, N.K. and Shukla, A.K. (2008) Use of some essential oils as post harvest botanical fungicides in the management of grey mould of grapes caused by Botrytis cinerea. World Journal of Microbiology and Biotechnology, 24, 39-46.
30. Kandhari, J. (2007) Management of sheath blight of rice through fungicides and botanicals. Indian Phytopathology, 60, 214-217.

31. Muthamilan, M. and Revathy, N. (2007) Management of tungro virus disease of rice with antagonists and botanicals. Journal of Ecobiology, 19, 81-88.
32. Sunder, S., Singh, R. and Dodan, D.S. (2005) Management of bacterial blight of rice with botanical extracts and non-conventional chemicals. Plant Disease Research, 20, 12-17.
33. Sitara, U., Naseem, J. and Sultana, N. (2008) Antifungal effect of essential oils on in vitro growth of pathogenic fungi. Pakistan Journal of Botany, 40, 409-414.
34. Arroyo, F.T., Moreno, J., Daza P., Boianova, L. and Romero, F. (2007) Antifungal activity of strawberry fruit volatile compounds against Colletotrichum acutatum. Journal of Agricultural and Food Chemistry, 14, 5701- 5707.
35. Vaughan, S.F., Spenser, G.F. and Shasha, B.F. (1993) Volatile compounds from raspberry and strawberry fruit inhibit postharvest decay fungi. Journal of Food Science, 58, 793-796.
36. Naidu, G.P. (1988) Antifungal activity in Codieum variegatum leaf extract. Current Science, 57, 502-503.
37. Harish, S., Saravankumar, D. Ebenezar, E.G. and Seetharaman, K. (2008) Use plant extracts and biocontrol agents for the management of brown spot disease in rice. BioControl, 53, 555-567.
38. Taiga, A. (2009) Efficacy of selected plant extracts in the control of fungal dry rot of white yam (Dioscorea Rotundata) Tubers in Kogi State. American-Eurasian Journal of Sustainable Agriculture, 3, 310-313.
39. Dorman, H.J.D. and Deans, S.G. (2000) Antimicrobial agents from plants: Antibacterial activity of plant volatile oils. Journal of Applied Microbiology, 88, 308-316.
40. Bajpai, V.K. and Kang, S.C. (2010) Antifungal activity of leaf essential oil and extract of Metaseqoia glystostroboides Miki ex Hu. Journal of the American Oil Chemists' Society, 87, 327-336.
41. Reddy, K.R.N., Reddy, C.S. and Muralidharan, K. (2009) Potential of botanicals and biocontrol agents on growth and aflotoxin production by Aspergillus flavus infecting rice grains. Food Control, 20, 173-178.
42. Enikuomehin, O.A. (2005) Cercospora leaf spot disease management in Seasame (Seasamum indicum L.) with plant extracts. Journal of Tropical Agriculture, 43, 19-23.
43. Abd-Alla, M.A., El Mohamedi, R.S.R. and Badeaa, R.I. (2006) Effect on some volatile compounds on black mould disease on onions bulbs

during storage. Research Journal of Agriculture and Biological Sciences, 2, 384- 390.

44. Kowalska, B. and Smolinska, U. (2008) The effect of selected plant materials and extract on the development of bacterial disease on onion. Vegetable Crops Research Bulletin, 68, 33-45.
45. Suleiman, M.N. and Emua, S.A. (2009) Efficacy of four plant extract in the control of root rot disease of cow pea (Vigna unguiculata Linn. Walp). African Journal of Biotechnology, 8, 3806-3808.
46. Choi, G.J., Jang, K.S., Kim, J.S., Lee, S.W., Cho, J.Y., Cho, K.Y. and Kim, J.C. (2004) In vivo fungal activity of 57 plant extracts against six plant pathogenic fungi. Plant Pathology Journal, 20, 184-191.
47. Hadizadeh, I., Peivastegan, B. and Hamzehzarghani, H. (2009) Antifungal activity of essential oil from some medicinal plants of Iran against Aletrnarea alternata. American Journal of Applied Sciences, 6, 857-861.
48. Satish, S., Raveesha, K.A. and Janardhana, G.R. (1999) Antibacterial activities of plant extract on phytopathogenic Xanthomonas compestris pathovars. Letters in Applied Microbiology, 28, 145-147.
49. Suprapta, D.N. and Khalimi, K. (2009) Efficacy of plant extract formulation to suppress stem rot disease on vanilla seedlings. ISSA Journal, 15, 34-41.
50. Okigbo, R.N. and Ogbonnaya, U.O. (2006) Antifungal effect of two tropical plants leaf extracts (Ossimum gratissimum and Aframomum meleguata) on post harvest yam (Diascorea spp.) rot. African Journal of Biotechnology, 5, 727-731.
51. Atawodi, S.E. and Atawodi, J.C. (2009) Azadirachta indica (neem) a plant of microbial and pharmacological activity. Phytochemistry Reviews, 8, 601-620.

4

CHAPTER

APPLICATION TECHNOLOGIES FOR ASIAN SOYBEAN RUST MANAGEMENT

Carlos Gilberto Raetano, Denise Tourino Rezende and Evandro Pereira Prado

São Paulo State University "Julio de Mesquita Filho", FCABO Brazil

INTRODUCTION

Occurrence

Soybean rust is a foliar disease which initially surfaced and remained for many years in Asian countries such as Taiwan, Thailand, Japan and India (Ozkan et al., 2006). After that, the disease was detected in Uganda and South Africa and more recently in South America. Asian Soybean Rust (ASR) is caused by the *Phakopsora pachyrhizi* Sydon & P. Sydon fungus and has been the worst disease in soybean culture. The disease has been present on the American continent, in Paraguay and southern regions of Brazil since 2001 (Yorinori et al., 2010). The importance of ASR disease in Brazil can be evaluated by its rapid expansion and severity and the subsequent economic losses. Over three years (2001 to 2003), ASR dispersed to all soybean producing regions of Brazil, reached the whole of the American continent and was detected in the United States of America in November, 2004 (Yorinori, 2010). ASR disease, when not controlled, can cause a total loss of production (Yorinori et al., 2004). In Brazil, crops free of disease can have an average productivity of 3,300 kg ha^{-1}. However, with the production cost included for a return, net profits of 2,436 kg ha^{-1} have been seen, thus it is recommendable to control the causal agent of the disease (Yorinori, 2005). In the 2007/08 season, ASR showed the lowest severity level since the 2002/03 season, due to farmer awareness of the necessity to obey the "period of sowing interruption", instituted by many of states of the Brazilian Federation. Another cause for improvement was the predominance in the planting of earlier varieties and the improved monitoring system of the disease (Yorinori et al., 2010).

Severity of ASR disease

The permanence of the pathogen inoculum in the fields the whole year around, due to the post-harvest sowing of the summer season and due to sowing carried out under irrigation during the winter/spring crop season, it was difficult to control the disease between the 2003 and 2005 seasons. During this control period, in the western region of Brazil, the first symptoms of rust were already visible 18 to 30 days after emergence began (V3/V4) and, therefore, some crops received up to seven applications of fungicides. In 2005, with the liberation of the genetically modified soybean culture Roundup Ready (RR), between harvests, the situation became even more serious due to the permanence of the pathogen in the fields all year round (Yorinori et al., 2010). With the "period of sowing interruption" hen only the cultivation of soybeans used in research and for increasing generations provided by breeding lines is permitted under severe rust control conditions and subject to government control organs (MAPA), the severity of the disease has diminished. Nevertheless, the use of control practices of low efficiency, with inadequate fungicides, the use of reduced doses for lowering costs and inadequate number and duration of applications, unfortunately, contribute to persistence of the disease resulting in significant production losses. Continuous monitoring programs, adequate handling practices and appropriate application technology are necessary in order to guarantee the production of soybean culture. The relationship between lateness in the control of ASR and the severity of the disease is 0.25% for each day in which control is not carried out. The relationship between the return rate of the soybean and the severity of ASR is of -36 kg ha^{-1} for each severity percentage point (Calaça, 2007).

Control of the pathogenic agent

In order to define the strategies to be used for ASR control, regarding application technology, there must be an awareness of the way systemic fungicides move into plants after application and absorption has been carried out. In the present-day market, the majority of fungicides recommended for ASR control move from the base to the top of each leaf, with little chance of moving in the other direction and without the possibility of dislocation from one leaf to another (Antuniassi, 2005). Amongst the fungicides available at present for pathogen control, the triazole fungicides, when used alone, have not presented good performance, as can be seen with ciproconazole, propiconazole and meticonazole (Yorinori et al., 2010). The consistency shown in programmes for chemical control applied in a curative and preventive manner on different soybean varieties and growth stages of the crop has been evaluated by Navarini et al. (2007). The authors established that there was a tendency that higher profit rates were related to preventive applications between the R1 and R3 stages. They also established a low efficiency rate in the control of the pathogen when the fungicide propiconazole was applied in a preventive manner. A deficiency in the control of *P. pachyrhizi* was also observed 30 days after the spraying of the fungicide difeconazole on this crop, in a comparative evaluation of fungicides carried out by Soares et al. (2004). It therefore, becomes evident that triazole fungicides have some limited systemic activity (moving through the plant, especially to newly developed leaves) and are thus somewhat forgiving if the application is less than perfect. When triazole fungicides are mixed with strobilurin fungicides, they show better performance in the control of rust disease. In Brazil, it is believed that the causes of control failures may be related to technical failures in application, predominantly in a population shift which is more tolerant to triazole fungicides in some regions (*Fungicide Resistance Action Committee – FRAC*), instead of developing a tolerance or resistance to the triazole fungicides through genetic mutation of the fungus. As a precautionary measure, class representatives of the producers recommend the use of triazoles only when mixed with other groups of fungicides.

Economic impact of ASR in Brazil

An estimate of the volume of grain losses and of the economic impact of ASR in the period between 2002 and 2009 reached 34.2 million tons, a value equivalent to more than half a full soybean harvest. On the other hand, the economic impact of ASR, adding up grain losses (US$ 7.95 billions), control costs (US$ 5.76 billions) and intake losses (US$ 1.55 billions) during the same period, totalled US$ 15.25 billion (Yorinori et al., 2010). If we consider that the world demand for soybeans is strongly linked to population increase, to world riches and to bioenergy, the necessity to minimise losses in all stages of soybean processing becomes evident. These demand indicators generate worries, particularly in Brazil, where the biofuel program has demanded a mixture of 4% biodiesel in the final fuel formulation since 2009, especially since the main source of biodiesel is soybean oil (Barros & Menegatti, 2010).

FUNGICIDE SPRAY COVERAGE

The right moment for application is determined by climate conditions, the presence and severity of the disease, plant growth stage and fungicide efficiency (Yorinori et al., 2004). These factors, together with correct calibration of the application equipment and with correct handling practices aimed at the control of *P. pachyrhizi*, have not been sufficient to impede the advance of the disease in soybean culture. The necessity for more efficient application of phyto-sanitary products has been related to various researchers such as Adam (1977), Matuo (1990) and Van De Zande et al. (1994), amongst others. It can therefore be noted that, in order to obtain the best efficiency, the study and development of new application technologies are indispensable. Phyto-sanitary products must be applied with maximum efficiency and, for this to occur, studies of spray deposition and coverage and spray drift are necessary. This last factor is responsible for losses and is also a cause of environmental contamination (Matthews, 1992).

Table 1. ASAE Standard S-572 Spray Quality Categories

Droplet category	Colour	Symbol	MVD (μm)
Very fine	Red	VF	<150
Fine	Orange	F	150-250
Medium	Yellow	M	250-350
Coarse	Blue	C	350-450
Very coarse	Green	VC	450-550
Extremely coarse	White	XC	>550

Michigan State University – Department of Plant Pathology, USA

Droplet size and spray coverage

A definition of droplet size and the volume to be applied must be a priority in the planning of an application. Further factors, such as the correct time of application, weather conditions, product recommendations and operational conditions, should be considered as a whole, looking towards maximum performance with the least losses and the least environmental impact (Antuniassi, 2010). Spray volume has the greatest impact on canopy penetration and leaf coverage. Increasing the volume improves penetration and coverage. The recommended spray volume differs for each fungicide. For aerial applications, the minimum recommended volume is 5-7 gallons per acre (47-65 L ha^{-1}). Recent research on soybean canopy coverage for ground applications at different growth stages of soybean (R1, R3 and R5) support recommendations that a spray volume of 15 gpa (140 L ha^{-1}) may provide adequate coverage of the entire canopy early in the growing season (R1 and R3) but 20 gpa (187 L ha^{-1}) is necessary later in the growing season (by R5) when the soybean canopy density and volume have increased (Brown-Rytlewski & Staton, 2010). In Brazil, the spray volume rates for conventional ground spraying of soybean have varied from 100 to 150 L ha^{-1}, but it is possible to have a reduction of 50% in spray volume using the new spray technologies and earlier varieties. In the mid-west region (Cerrado), the use of low application rates with conventional ground sprayers is limited by climatic conditions due to the high temperature (30 to 40°C) and low air humidity (12 to 30%) during the greatest part of the year. Droplet size is the second most important factor affecting canopy penetration and leaf coverage. Research has shown that fine to medium droplets, with median volume diameters (MVD) in the range of 200 to 350 μm, maximise canopy penetration and leaf coverage. Smaller droplets provide better leaf coverage but lack the momentum to penetrate the canopy. Larger droplets have the momentum to penetrate the canopy but do not provide sufficient leaf coverage. Ground speed, nozzle pressure and spray volume should beconsidered when selecting nozzles for the sprayer. Choose nozzles that will produce 200-350 μm droplets at 15 to 20 gallons per acre (140 to 187 L ha^{-1}) while travelling at the desired speed. In most cases, nozzles for herbicide applications should not be used for fungicide applications as they are designed to generate larger droplets at lower application rates. All nozzle manufacturers use a spray classification system (ASAE standard S-572) of six categories with corresponding colours to classify the droplet size range produced by nozzles under various operating pressures. The colour of the nozzle itself should not be confused with the colours listed in Table 1. The nozzle colour describes the flow rate for the nozzle and the colours on the table describe the nozzle's droplet size range. When using droplet size classification charts, select nozzles that produce droplets near the fine end of the medium (yellow) category.

Ground speed affects spray volume and vertical droplet velocity. Taking into consideration that in order to apply fungicides, a fine to medium category of drops are indicated, and that the maximum wind speed during spraying should not surpass 9.6 km h^{-1} (Andef, 2004), a critical new situation presents itself in the field. Auto-propelled sprayers present innovations that give greater stability to the spray booms and with this, the operational speed increases to values rearing and even above 16 km h^{-1}. The immediate consequence of this operational situation is that the relative wind between the boom in displacement and the air canopy which is present between the spray boom and the intended crop have a braking effect, contrary to the downward speed of the fine droplets generated at the tips of the sprayers. This process help with evaporation and also with the drift of the fine spray droplets and hinders its arrival on the crop canopies to be treated. A second consequence depends on middle-sized droplets that manage to maintain their falling speed in spite of the opposite effect generated by the dislocation speed of the boom. Research carried out recently on winter cereals, by the Institute DLG in Germany, shows that these droplets deposit themselves, on the whole, only on one side of the plants, with the other side ("shady side") consistently lacking in droplets (Boller & Raetano, 2011). The research also revealed that an increase in the displacement speed of the equipment implies in a greater deposit of droplets on the upper third of the plants and fewer droplets deposited on the lower leaves. The increase in spraying pressure may partially compensate for this effect; however; one cannot emphasise too strongly that excessive working pressure is one of the most important factors that facilitate spray droplet drift. This picture deserves particular attention, due to the fact that the actual and future tendency is the increase in the displacement speed of the spray equipment by land. In the same situation, spray nozzles with flat double spray outlets show a slight increase in the quantity of droplets deposited on the side known as "the shad side". The most balanced situation was obtained when ends with flat double jets, with differentiated angles in relation to the vertical position, were utilised. The results indicate that this type of outlet may be efficient for a more even deposition of the droplets, on both sides of the plants, when the displacement speed of the boom is around 12 km h^{-1} (Boller & Raetano, 2011). There are basically two ways to increase coverage: 1) reduce droplet size and 2) increase carrier volume (application rate). Large droplets do not provide good coverage and result in chemical wastage. Increasing the application rate may be equally undesirable. It requires frequent refilling of the sprayer tank. This wastes time that may be extremely valuable when there is a short period of opportunity to spray. Ideally, we want to have as many small droplets on the target as possible. However, extremely small droplets have a tendency to drift. Research has shown that there is a rapid decrease in the drift potential of droplets whose diameters are greater than approximately 200 μm. When extremely small droplets are released from the nozzle, they quickly lose the momentum that is needed to push the droplets into the canopy. Also, these extremely small droplets do not last long after they are released from the nozzle. Most of them evaporate within a few seconds (Ozkan, 2010). The single most important factor affecting the control of ASR disease is to get a thorough coverage of soybeans with the fungicide, which is much different and more challenging than spraying for weeds and insects. The most effective coverage on soybean plants can be obtained with both the horizontal as well as vertical distribution of the fungicide on soybean leaves. Asian soybean rust usually shows its symptoms in the lower parts of the plant first and works itself up towards the top of the plant. The most effective spray equipment and methods for applying fungicides on soybean plants to control Asian soybean rust was studied by Ozkan et al. (2006). A second component of the study was to determine the effect of spray quality (fine, medium, coarse) on spray deposition and coverage using three different sizes (8002, 8004 and 8005) of the XR type of a flat fan nozzle operated at different spray pressures. The application rate was kept constant at 145 L ha^{-1} for all the treatments. The average spray coverage on the middle part of the soybean canopy (0.6 m above the ground) varied from 1.3 to 7.3% among the treatments. The Jacto sprayer provided the highest spray coverage on the middle part of the canopy, followed by Top Air sprayer and the boom sprayer with a TX-18 hollow cone nozzle that produced the lowest spray coverage on the middle part of the canopy, followed by Turbo duo, and then XR 8002 nozzles. The average spray coverage at the bottom part of the soybean canopy (0.3 m above the ground) varied from 0.5 to 3.9% among the treatments. Similarly to the coverage on the middle part of the canopy, the Jacto sprayer provided the highest spray coverage on the bottom part of the canopy, followed by the boom sprayer with the canopy opener and then the Top Air sprayer. The boom sprayer with XR 8002 nozzles produced the lowest spray coverage on the boom part of the canopy, followed by hollow cone TX-18 nozzles. XR 8002 flat fan nozzles and hollow cone nozzles had smaller MVD than other treatments with the boom sprayer. The authors observed that among the three spray qualities (fine, medium and coarse), the medium quality spray provided the highest coverage and the fine quality spray provided the lowest coverage at both middle and bottom parts of the canopy. When compared to the XR 8004 flat fan pattern nozzles with medium spray quality, Twinjet, Turbo dual pattern nozzles and hollow cone nozzles provided very low coverage on the middle and bottom parts of the canopy. Droplets from Twinjet, turbo dual pattern and hollow cone nozzles had poor penetration capabilities because these droplets had horizontal velocities. The horizontal movement of droplets consumed kinetic energy and caused droplets to easily settle on the top leaves. The influence of the size of droplets from different nozzles on soybean spray coverage was studied by Antuniassi et al. (2004). The authors verified that very fine quality spray obtained with hollow cone TX VK6 nozzle and Twinjet flat fan TJ 60 11002 nozzle, and fine quality spray with a flat fan pattern XR 11002 nozzle, provided greater coverage in middle and bottom parts of the soybean plants when compared to the extremely coarse spray quality produced by air induction flat fan nozzles. The effects of spray nozzles (flat fan pattern, pre-orifice flat fan, air induction flat fan and air induction twin flat fan) and volume rates (115 and 160 L ha^{-1}) on chemical control of rust and the deposition of tebuconazole fungicide sprayed on soybeans of the Emgopa 313 variety, were studied by Cunha et al. (2006). The results showed that, despite the fact of the volume rate of 160 L ha^{-1} and of the use of pattern flat fan nozzles, they provided larger fungicide distribution uniformity in the plant canopy. There was no influence of the nozzle type neither of the application volume in the control of the rust, as well as in the soybean yield. In part, the results described by Raetano & Merlin (2006) ratified those observations that have been made by Cunha et al. (2006). The experiments were conducted in 2004/05 and 2005/06 seasons, using soybean, IAC-19 variety, with the same sprayer equipment and near application volumes (99 and 143 L ha^{-1}; 100 and 150 L ha^{-1}). The values of spray deposition were less influenced by nozzle type (hollow cone, flat fan and twin flat fan), both with fine spray quality. It is recommended for Asian soybean rust control that droplets have a size of 200 to 300 μm (OZKAN, 2005), but droplets smaller than 100 μm can be used with drift control in spraying with air assistance delivery systems near to the sleeve boom.

FUNGICIDE APPLICATION TECHNIQUES

Nowadays, Asian soybean rust (ASR) deserved special attention due to its severity and difficulty of control, since it develops in the aerial part of plants, damaging the physiology and contributing to a drastic reduction of grain yield. For efficient control and cost-cutting, spray techniques and spray equipment must be improved. Studies show that the use of air assistance in the sleeve boom, connected to the hydraulic system of the tractor, can reduce the drift, increase droplet penetration into the plant canopy and improve the spraying distribution (Bauer & Raetano, 2000; Cooke et al., 1990; Taylor et al., 1989; Taylor & Andersen, 1991).

Air assistance delivery system in boom sprayers

The use of air assistance in phyto-sanitary product application is very old. However, the enthusiasm in using this spray technology started in 1980, as reported by Robinson (1993). Four years later, the Degania Sprayers Company in Israel developed a sprayer, revolutionary at the time, equipped with air assistance on the spraying sleeve boom. However, only since the end of the 1980s and the beginning of the 1990s has air assistance been effectively adopted in sleeve boom sprayers. In Europe, this technology was introduced by Hardi, and in Germany, in 1996, seven manufacturers exhibited equipment with air assistance in the *Agritechnica* agricultural trade show (Koch, 1997). At that time, the Brazilian industry also incorporated this technology to tractor-driven trailing sleeve boom sprayers. The incorporation of this technology to sleeve boom sprayers was an attempt to improve spraying penetration in the target culture, reduce drift and the number of applications required, increase the time available for carrying out the spraying and enable changes to the spraying height over the culture (DEGANIA SPRAYERS Co., s.d.). Forapplying phyto-sanitary products on low-stem cultivation, the spraying sleeve booms equipped with air assistance appeared as the ideal tools to improve application quality (smaller droplets, in higher numbers), increase productivity (lower volume and replenishment, higher displacement speed and extended spraying times), reduce drift (the machine's wind speed is greater than the environmental wind) and exposure to these products (Sartori, 1997). After twenty years of using air assistance in sleeve boom sprayers, a great deal of information must still be clarified about the interactions between air volume and speed which are more appropriate for different cultures, the angle of the nozzles on the boom in relation to the air, spraying height and displacement speed, amongst other factors which enable wider spraying coverage and lower losses.

Characterisation of the technology

Tractor-driven sprayers with air assistance can be coupled to the tractor's hydraulic power take-off (third point) (those with lower capacity tanks or of the trailing type). These sprayers are equipped with one or two fans, usually axial, positioned near the centre section of the spraying sleeve boom, which distribute a very high air volume in an inflated duct assembled over the boom and nozzles (Matthews, 2000). The speed of the air generated may vary with the fan rotation (rpm), and generally it does not follow a linear relationship. Also, air speed variations could occur along the boom, at the ends, when compared to the speed achieved in its centre section (Raetano, 2002). The established standards for evaluating with accuracy the speed of the air generated by sleeve boom sprayers equipped with air assistance was necessary to standardise the measuring distance in relation to the air exit opening, as well as to specify anemometers that are able to record high air speeds (30-40 m s^{-1}). Thus, Kunz (2010) developed two methods for air speed and volume measuring in spray booms equipped with air assistance. In the first method, a wooden mould was placed at the outlet of the air curtain, in a vertical position, in the direction of the air flow, and measurements were taken with the anemometer at pre-established distances. This form of measuring became know as the "ruler method". This method makes it very difficult to determine the main vector of the air flow that comes out in a continuous manner through the rectangular opening on the lower part of the inflated sleeve, which makes it difficult to measure the air speed with precision. New air-speed readings are now taken beforehand, using a nylon thread fixed to the air outlet, to indicate the point of air flow displacement vector, which substitutes the ruler method. In this way, it makes it very much easier to identify the main air flow and increases the precision and uniformity of speed values obtained with the anemometer. In a similar manner to the ruler method, pre-defined distances are marked off on the nylon thread, so that measurements can be taken with greater ease and accuracy. This procedure is called the "thread method" (Figure 1). Due to the dynamic behaviour of the air flow, it becomes difficult to identify the vector of the air flow that comes out under high speed from the system, principally at distances of 0.25 and 0.50 m, which causes a great variation in the speed data obtained with the ruler method, as can be seen in Table 2. The air speed values obtained with the thread method present greater uniformity in relation to the ones obtained with the ruler method, especially at distances of 0.25 and 0.50 m from the air outlet. This can be observed through the variance values (%) of the data, which were smaller with the thread method (Table 2). The average values of the air speed obtained with the thread method were greater, probably due to the correct identification of the main air flow vector when measured by this method. Measuring the air speed, therefore becomes more precise and easier, especially at longer distances in relation to the air flow at the spray boom.

Fig. 1. "Nylon thread method" for measuring air speed in a sleeve boom sprayer.

Table 2. Descriptive statistics of the air speed data obtained along the spray boom with different evaluation methods.

Distance (m)	Average*	S.D.	Variance	CV %	Min*	Max*	Amplitude*
Ruler method of measurement							
0.0	70.14	10.00	100.05	14.26	53.40	97.20	43.80
0.25	41.71	5.70	32.55	13.68	31.20	54.00	22.80
0.50	29.49	8.20	67.27	27.81	20.50	51.80	31.30
Nylon thread method of measurement							
0.0	71.57	10.09	101.83	14.10	59.50	93.60	34.10
0.25	43.64	4.04	16.33	9.26	36.60	51.00	14.40
0.50	35.26	3.53	12.47	10.01	28.80	41.90	13.10

*Values expressed in km h^{-1}.

1.1.1 Air speed on spray deposition

Raetano & Bauer (2003) evaluated the effects from air speed variation (50%, 75% and 100% of the maximum fan rotation capacity) on the spraying sleeve boom, when depositing phyto-sanitary products on bean culture. Forty-eight days after sprouting begins, 200 g of copper oxide per 100 L of water were applied with AXI-110015 tips at 206.7 kPa and JA-1 at 1,033.5 kPa, either with air assistance or not, using a

Model Falcon vortex sprayer. The broth volume was 100 L ha^{-1} in both operational conditions. The air speed variation did not influence the deposit levels in the culture, but the use of air assistance, operated at full fan capacity, resulted in better deposit levels on the abaxial surface of the leaflets positioned in the lower portion of the plants. Cereal-cultivated soil contamination can be reduced to approximately 40% when using 50% of the maximum speed of the air generated by the fan in a sprayer equipped with air assistance on the sleeve boom, when compared with conventional application (without air), as reported by Taylor & Andersen (1997). The deposit and losses of spraying broth in the cultivation of bean (*Phaseolus vulgaris*), 26 days after sprouting, and using sprayers equipped with air assistance on the sleeve boom and conventional sprayers (without air) and volumes of 60 and 100 L ha^{-1}, have been evaluated by Raetano & Bauer (2004). The higher volume resulted in greater deposits, but high losses to the soil (above 60%) have been noted, even when using air assistance with air speed corresponding to 50% of the maximum fan rotation. In part, such results have been assigned to 40% of the soil bare of vegetation at this growth stage of the culture. The air volume generated may vary from 0 to 2000 m^3 per hour per boom, depending on the number and power of the fans distributed on variable-size booms that could reach 30 m in length. The air distributed in the inflated duct is forced to pass through a continuous or intercalated opening, in a perpendicular direction to the one in which it has been generated, in a descending direction. The effects of chemical control of the rust and deposition fungicide sprayed under four speeds (zero, 9, 11 and 29 km h^{-1}) by a spray boom on soybean crop were evaluated by Christovam (2008) and Prado et al. (2010). Significant differences were obtained in the lower part of the plants for spray deposition using higher speed of air assistance. On the top part of the plants, greater levels of deposition were seen when spraying without air assistance was carried out. The rust severity was more intense in treatments without air assistance. Raetano & Bauer (2003) evaluated different velocities of air assistance near the spray boom and concluded that air assistance, with maximum air speed generated by the fan (29 km h^{-1}) and a flat fan nozzle (AXI 110015 type), provided greater spray deposition on the abaxial leaf surface, on the bottom part of the bean plants. The data of the soybean crop yield at different air speeds using an air-assisted sprayer for Asian soybean rust management was compared in the 2006/07 (Christovam et al., 2010a) and 2007/08 (Prado et al., 2010) seasons (Figure 2). Air speeds used in both studies were zero, 9, 11 and 29 km h^{-1}.

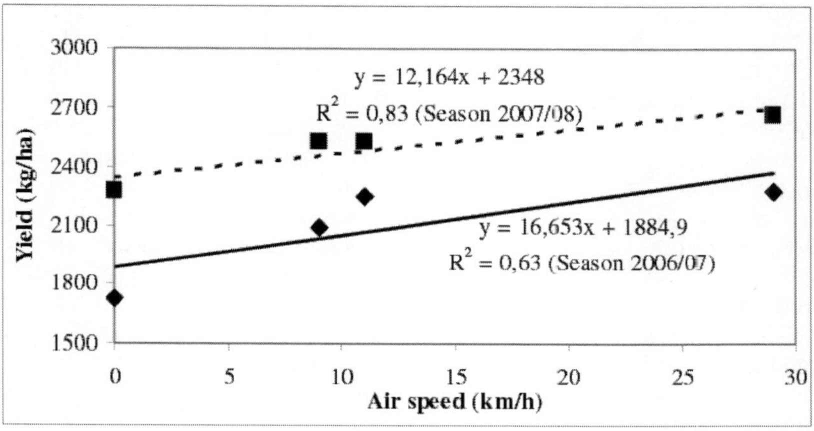

Fig. 2. Effect of the air speed on soybean crop yield, 2006/07 and 2007/08 agricultural

There is a positive correlation between air speed and soybean crop yield. When compared, the soybean yield using the maximum air speed generated by the fan (29 km h^{-1}) in conventional spraying (without air), increases of 31.9% and 17.1% can be seen in the 2006/07 and 2007/08 seasons, respectively. As can be verified, in the last agricultural season, the increase in soybean yield was lower, due the higher severity of ASR on the Conquista variety (Figure 2). The effect of different air speeds (0, 9, 11 and 29 km h^{-1}) in chemical control of pests on soybean crop, Conquista variety, was evaluated by Prado (2009), after insecticide spraying using an air-assisted sprayer. The use of air speed generated by the fan in the maximum rotation (29 km h^{-1}) provided greater control of the lower velvetbean caterpillar (*Anticarsia gemmatalis*). This effect was not observed with longer caterpillars (> 1.5 cm) because the lowest caterpillars are more easily located in the lower part of the soybean plants. Thus, the lower caterpillars received an additional amount of insecticides due the effect of air assistance with maximum air speed into the canopy. In general, there was not a statistically significant difference between air speeds on stink bug control after insecticide spraying on soybean culture.

Nozzle angle on spray deposition

The positioning angle of the spraying nozzle in relation to the air curtain (Figure 2), generated by the equipment (vertical, descending), as well as the nozzles and air curtain, simultaneously, in relation to the vertical position, may significantly influence the deposit levels and the spraying distribution. Nowadays, in sleeve boom sprayers equipped with air assistance, the angle variations of the nozzles and air curtain, in relation to the vertical position, pro or against the tractor-sprayer assembly displacement, are made simultaneously clockwise or counterclockwise, with the single-cylinder command. The results research carried out under controlled conditions and in the field have shown that the positioning of the nozzle at 30° forward of the displacement in conventional sprayers (without air) provides a significant increase in deposits on the leaf surface of different vegetal species: *Cyperus rotundus* (Silva, 2001), *Brachiaria plantaginea* (Tomazela, 2006) and *Glycine max* (Bauer, 2002). In England, research carried out in wind tunnels with plants cultivated on trays have confirmed that the spraying angle forward of the displacement, in the presence of air assistance, increased deposit on cereals and reduced soil contamination (Hislop et al., 1995). Nowadays, one may position the spraying nozzles and air curtain at angles of 15° and 30° in relation to the vertical position in sleeve boom sprayers equipped with air assistance, made in Brazil. The use of air angled forward of the displacement with fine droplets could substantially increase spraying deposit levels on vertical targets. These results were obtained from practical experiences published by the Hardi Int. Tech. Reports in potato culture which indicated that spraying penetration and retention are greater with air assistance positioned at an angle forward of the displacement on the leaves in the lower portion of the plants. In the upper portion, the retained broth volume was virtually not influenced by the air exit angle, pro or against the equipment displacement (Taylor & Andersen, 1997). The effect of the nozzle angle and air-jet parameters in an air-assistance sprayer on the biological effects of ASR chemical protection was studied by Christovam et al. (2010b). Four air levels (0, 9, 11 and 29 km h^{-1}) were combined at two nozzle angles 0° and 30° for the sprayings using flat fan AXI 110015 nozzles. The spraying with triazole fungicide was realised in R2 and R5.2 growth stages of soybean at 142 L ha^{-1} of volume rate. For the evaluation of spray deposition, a cupric tracer was used. At the bottom part of the plant, spraying with maximum air speed generated by the fan and nozzles angled at 30°, it was essential to promote doubled deposits on the abaxial leaf surface (Table 3). Maximum air speed (29 km h^{-1}) and nozzles angled at 30° resulted in an increase in spray deposits on adaxial surface of leaves in the bottom part of the plants (Table 3).

Table 3. Average values of deposits of the copper tracer in an artificial target (filter paper) on leaf surfaces, in the bottom part of the soybean plants, Conquista variety, in relation to different spraying angles. Botucatu, SP, 2006/2007.

Air Speed ($km\ h^{-1}$)	Adaxial surface				Abaxial surface			
	Angle 0°		Angle +30°		Angle 0°		Angle +30°	
	$\mu l\ cm^{-2}$		$\mu l\ cm^{-2}$		$\mu l\ cm^{-2}$		$\mu l\ cm^{-2}$	
0	1.2425	a A	0.6962	b AB	0.6585	a A	0.2444	b B
9	0.6997	a A	0.9865	a AB	0.3621	a A	0.4527	a B
11	1.147	a A	0.6395	b B	0.4651	a A	0.3292	a B
29	0.6287	b A	1.2663	a A	0.4904	b A	0.8552	a A
DMS Angle	0.46				0.24			
DMS Air speed	0.62				0.33			
CV (%)	34.17				34.16			

The same larger letters in the column, did not differ by the Tukey test ($p<0.05$).

Nozzles angled at 30°, in the same direction of the sprayer displacement, combined with air assistance (29 km h^{-1} of air speed) positively influenced the control of disease as well as the yield of the Conquista variety crop. This fact confirms the importance of spraying performed with nozzle angles in the same direction of the sprayer movement, which can contribute significantly to Asian soybean rust control, considering the disease epidemiology. The choice of the best combination of air speed and nozzle angle in air-assisted sprayers is influenced by architecture and growth stage of the plants to obtain a desirable biological effect in soybean Asian rust chemical protection with this technology. Conventional spraying (without air) and air-assistance at 0º (vertical) and 30º (forward to displacement of the equipment) are shown in Figures 3A, 3B and 3C, respectively. The spray boom angle interference, with or without air assistance near the boom, on spray deposit levels were studied by Scudeler & Raetano (2004) in potato culture. The higher deposits were evidenced with nozzles positioned at 0º and 30º, with the presence of air assistance, both at the top and bottom part of the potato plants. The lower spray deposits were obtained with nozzles positioned at 30º in the opposite direction to the displacement of the sprayer. In addition to the volume rate, generated air speed and nozzle angle in air-assisted sprayers, other factors, such as displacement speed of the tractor-sprayer assembly, presence of vegetal coverage in the area or not, vegetal coverage type (monocotyledonous or dicotyledonous, plant density, architecture and plant cuticle characteristics), positions of insect pests and plant pathogens, agrochemical product characteristics, droplet size and environmental conditions, especially wind speed, may influence the efficacy of phyto-sanitary control. It is necessary to develop studies with variations in air speed combined at different angles of spray nozzle on spray deposition and coverage. Dynamic systems for air speed evaluation combined at different nozzle angles and the performance of these in spraying could be better studied.

Figure 3. Air-assisted sleeve boom sprayer in the following operation modes: A – conventional spraying (without air); B – spraying with air assistance at 0º (vertical) and C – spraying with air assistance angled at 30º forward to the displacement of the tractor-sprayer assembly on soybean crop.

Effect of the air-assistance delivery system on soybean productivity

The influence of an air-assisted delivery system on soybean productivity, var. Conquista, 2006/07 and 2007/08 seasons are shown in Figure 4. The spraying treatments used to control *P. pachyrhizi* fungus were applied on soybean plants (R2 and R5.2 growth stages) using a triazole + strobilurin spray mixture and volume rates between 120 at 150 L ha^{-1}. The data of soybean crop productivity after two sprayings with different technologies was submitted to a variance analysis and the averages were compared by Tukey's test ($p<0.05$).

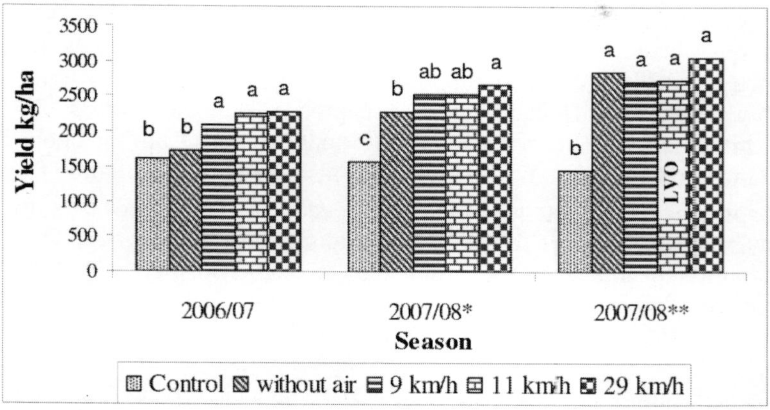

Figure 4. Effect of treatments: control (not treated); air-assistance delivery system at zero (without air), 9, 11 and 29 km h^{-1} air speed; rotating nozzle-LVO with a third part of spray volume (40L ha^{-1}) on soybean productivity. The Conquista variety was used in the 2006/07 and 2007/08 agricultural seasons. Botucatu, SP, Brazil (2006/07 - Christovam et al. 2010b; 2007/08* - Prado et al. 2010; 2007/08** - Christovam et al. 2010a).

In general, this technology provided higher productivity, in relation to that with conventional spraying (without air) and control (not treated). A higher increase in soybean productivity was obtained with maximum air speed generated by the fan (29 km h^{-1}). Theair assistance constituted an optional implement on the boom sprayers that can increase up 30% the equipment cost in relation to conventional sprayers which do not have this technology. Although the use of this technology should not be recommended on soils without vegetation or even on early stages due to the smaller foliar area. This method provides several advantages that have been mentioned before.

Air assistance and spray drift

Air assistance in the spraying sleeve boom significantly improves the penetration of spraying, especially in high cultures and high leaf density, such as potato, in addition to reducing drift (Koch, 1997). However, these effects were not observed when air-assisted spraying is done on bare soil or plants in the first stages of growth. Also, according to Matthews (2000), air-assisted spraying penetration is better when compared to conventional spraying on wide-leaf cultures, such as cotton. In Holland, tests with the air-assisted sprayer Twin (Hardi) have been carried out on potato cultivation. Generally, air assistance reduced drift by sedimentation by 50% and the air-carried drift by 75%. In Holland, the accepted drift percentage by sedimentation is 8-10% for a distance of 1.5 to 2.0 metres from the boom, and around 0.2% for 5.0 to 6.0 metres intervals. The recommendation for making spraying in Holland is with a wind speed less than 5.0 m sec^{-1}. In Germany, the accepted drift values by sedimentation when applying phyto-sanitary products range from 0.6 to 0.1%, respectively, for distances of 5.0 to 30.0 metres from the spraying sleeve boom (Jorgensen & Witt, 2000). Considering the drift limits accepted for spraying in Germany, the safe distance for applying near water channels (irrigation/draining) in that country is 10.0 metres for 80% of the herbicides approved for use, and 20.0 metres for other herbicides. France and Belgium comply with the drift limits accepted in Germany. Artificial targets have been also used by the Morley Research Centre for simulating venomous plants in the sugar beet. The variations in the deposit values for air-assisted spraying were lower when compared to those achieved with the conventional sprayer (Taylor & Andersen, 1997). These authors have also demonstrated the influence of air assistance on drift percentage reduction compared with conventional application (without air), by obtaining 90, 84, 83, 76, 68 and 61%, respectively, when spraying barley, bean, pea, Brussels sprouts, lettuce and leek, with fine droplets. Nowadays, studies involving computer models aim at clarifying the relationship among the air released drift risk and deposit on target. Preliminary studies have shown that the increase of the displacement speed with air-assisted sprayers may reduce drift, but provide lower evenness in the target culture treatment (Miller, 1997). However, aiming at reducing the application volume, Nordbo (1992) has demonstrated lower variation and improved deposit by using air assistance. The density, architecture, cuticle type (pilose, glabrous and waxy) and growth stage of the vegetal species in the area are factors influencing phyto-sanitary control efficiency when using air-assisted sleeve boom sprayers. Fine droplets provide larger deposits on plants, especially monocotyledons, but are very susceptible to drift. Their penetration capacity in cultures is small, and then the loss to the soil must be limited. Therefore, air assistance enables using fine droplets more efficiently, by reducing the drift and increasing the deposits on the target, in addition to providing higher penetration of these droplets in cultures with higher leaf density, and reducing losses to the soil (Jorgensen & Witt, 1997). On the other hand, coarse droplets generally provide good drift control. In dicotyledons, the deposits do not depend only on droplet size (Nordbo,

1992). Unlike the results with smaller diameter droplets, coarse droplets provide significantly lower deposits on vertical surfaces (monocotyledons), and especially in the first growth stages, by increasing the loss to soil proportionally to their size (Jorgensen & Witt, 1997). In vegetables, where droplet retention is limited by the presence of waxy layers on the cuticle, further studies are required, especially with air-assisted spraying, in order to evaluate the application quality (Koch, 1997). In the absence of vegetation (bare soil), air assistance may increase drift and deflect the air from the sprayer by the soil, unlike the effect which occurs in the presence of vegetation, with the impact of droplets on the leaf surface (Matthews, 2000).

Figure 5. Canopy opener artefact fixed to the spray boom in a soybean crop.

Alternative Spray Technologies on Soybean

Nowadays, other technologies are available to the boom sprayers enabling higher spraying droplet canopy penetration in soybean culture. The difficulty in controlling Asian soybean rust and late season diseases has favoured the development of new spraying techniques, particularly due to the difficulty in reaching the exact target to be controlled. Thus, the use of the opener, rotating system nozzles, hose drops and electrification of droplets associated with air assistance can be mentioned.

Opener

Conventional sprayers linked to an artefact providing the canopy opener at the same spraying way can turn out to be an economic and effective alternative to soybean growers with lower purchasing power (Zhu et al., 2008b). These authors found that spraying performed with conventional sprayers linked to a canopy opener did not results significant differences in the coverage of the spray in the middle part of soybean plants when compared to spraying carried out using air assistance. However, the canopy opener coverage and air assistance along the bar was higher compared to treatments where the spraying was conducted by the conventional system without the canopy opener. Thus, the opener and spray boom coupling can provide deposition results similar to those obtained with the use of air assistance, besides being a more economical alternative to Asian soybean rust control (Figure 5). Considering the difficulties in controlling ASR by fungicide spraying, Prado (2011) evaluated the effectiveness of the canopy opener compared to conventional sprayers and air assistance in the spray boom on spraying deposition, rust control efficiency and soybean productivity. The experiment was conducted in a randomised block experimental design with six treatments: conventional spraying (T1), spraying with air assistance at maximum capacity of the fan rotation in the boom (T2), spray with a canopy opener to a depth of 0.10 m (T3), spraying with a canopy opener to a depth of 0.10 m with air assistance (T4), spraying with a canopy opener to a depth of 0.20 m (T5) spraying with a canopy opener to a depth of 0.20 m with air assistance (T6) and control treatment (without spraying) (T7) in four replicates, totalling 28 plots. The area of each plot was equivalent to 70 m^2. The depths of 0.10 and 0.20 m refer to the distance from the canopy opener in relation to the top of the soybean plant. In addition to the distance between the canopy depth opener and the top of the plants, there is also a predetermined horizontal distance of 0.15 m between the boom and the canopy opener. The function of the canopy opener is to promote the soybean plants to slope forward, opening a space in the plant canopy and thus facilitating the flow, and consequently droplet deposition, on the bottom of the soybean plants. The sprayer used in the experiment was the Advance Vortex model 2000 with an 18.5 m long boom, 37 flat fan XR 8002 nozzles spaced 0.50 m apart operating at a pressure of 295 kPa and a spray volume of 150 L ha^{-1}. The comparative effect of these different technologies on soybean productivity after three fungicide pyraclostrobin + epoxiconazole mixture sprayings at a dose of 25 + 66.5 g a.i ha^{-1} in the development stages R2, R3 and R5, as shown in Figure 6. The treatment T5 (spraying with a canopy opener at a depth of 0.20 m) had a higher productivity increase (54%) compared to control treatment. All treatments which received fungicide had significantly higher yields than the control treatment. There was no difference between the canopy opener and air assistance on soybean yield, making it an interesting and economical alternative for the control of Asian soybean rust. These results corroborate those obtained by Zhu et al. (2008).

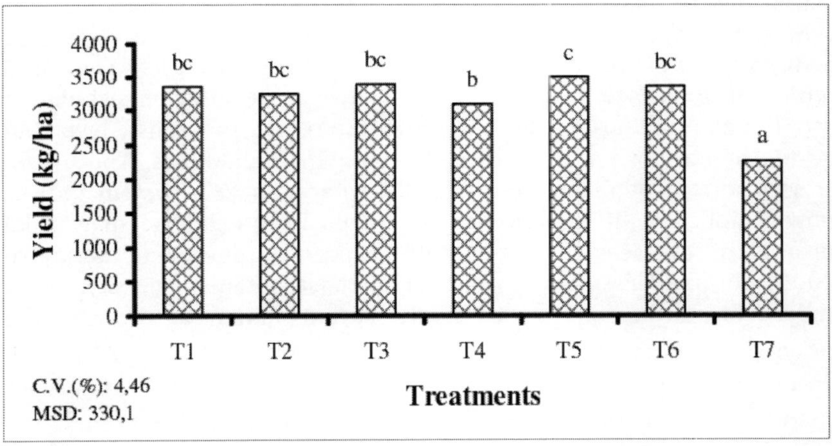

Figure 6. Effect of different treatments: conventional spraying (T1); air-assistance (T2); *canopy opener* at a depth of 0.10 m (T3); *canopy opener* at a depth of 0.10 m with air assistance (T4); *canopy opener* at a depth of 0.20 m (T5); *canopy opener* at a depth of 0.20 m with air assistance (T6) and treatment control (without spraying) on soybean yield.

Rotating system nozzles

Recently, the use of centrifugal energy to produce spray droplets (rotating system nozzle) is an interesting alternative application technology to control Asian soybean rust, using oily formulations, low spraying volumes and, consequently, a greater operational performance of sprayers. In Brazil, new techniques for pesticide application using low volumes and rotating nozzles have been developed in the mid-west region (Cerrado) for soybean rust control in soybean culture. The rotating nozzle low volume oily (LVO) and different levels of air speed with an air-assisted sprayer on spray deposits were compared by Christovam et al. (2010c) on Asian soybean rust control and soybean productivity. Two experiments were carried out in the experimental area of FCA/UNESP, Botucatu, SP, Brazil, on a soybean crop of the Conquista variety, in the 2007/2008 season. In the first experiment, three air levels (0, 9 and 29 km h^{-1} air speed generated by a fan) with flat fan XR 8002 nozzles and a spray volume of 130 L ha^{-1} were compared with a rotating nozzle – using LVO at 40 L ha^{-1} of spray volume (Figure 7). The second experiment was carried out under the same conditions as the previous experiment, including a control treatment (untreated plants). The grades varied between 0.6 and 78.5% disease severity. In general, air assistance promoted the increase on deposit levels on the adaxial surface of the leaf located in the top part of the plants. Therefore, in the bottom part, there was not a significant difference in spray deposits between the spraying techniques. Also, the abaxial surface did not show differences in deposit levels, in the top or bottom part, between the spraying techniques. The use of air assistance, when compared with the rotating nozzle system, did not show significant differences in spray deposits on adaxial or abaxial surfaces of the leaves in the bottom part of the plant. Monteiro (2006) observed results very similar to those obtained in the current study. This author performed a study that aimed to evaluate the spraying efficiency of a rotating atomiser system - LVO using 25 L ha^{-1} of fungicide outflow on a soybean crop, when compared to a sprayer equipped with hydraulic nozzles at a spray volume of 150 L ha^{-1}. The treatments sprayed with the fungicidal mixture provided a weight of 1000 seeds and productivity significantly higher in comparison with untreated plants (control). The highest increase of productivity was obtained with the maximum air speed generated by the fan (29 km h^{-1}) near to the spray boom using 130 L ha^{-1} when compared with the control treatment. The spray volume applied with the rotating system nozzle – LVO was 40 L ha^{-1}. Therefore, it did not provide the same increase in productivity compared with the treatment using air assistance at the maximum speed. The rotating system nozzle was 30% more economical than the treatment with a spray volume of 130 L ha^{-1}, with or without air assistance near the boom, using the Advance Vortex 2000 sprayer.

Hose drops

Another possibility to improve spraying coverage with boom sprayers is addressed by spraying with three flat fan nozzles involving the entire planting row, of which two of them are positioned near the bottom of the plants or positioned on opposite sides between the crop rows and near the bottom of the plant. The structures that support the spray nozzle from the spray boom at its lower end are called hose drops. In the USA, there are several reports on the use of flat fan nozzles placed in the hose drops ends, which move in the line between the culture, with volumes around 140 L ha^{-1} in Asian soybean rust treatment (Ozkan, 2005), although their culture is planted at greater spacing than those in Brazil with early cultivars. In Brazil, growers with difficulty will adopt hose drops in this application in order to obtain better spray coverage of the leaves on the bottom part of soybean plants. However, the differences in the growth habits, foliage degree and plant architecture of the varieties and the smaller spacing between planting rows makes the use of this technology difficult.

Electric charge (electrostatic) in spray droplets

Nowadays, air assistance can be combined with electrification (by induction) of the spraying droplets, aiming at reducing drift and exposure of appliers and the environment to phyto- sanitary products. An experiment was carried out in commercial areas of the soybean crop, Cidade Verde Farming, Primavera do Leste, MT, Brazil, on soybean plants of the Monsoy 8757 variety in the 2009/2010 agricultural season. Sowing was performed in 12/11/2009, leaving 0.45 m spacing between planting rows and 14 seeds per linear metre. The experimental design was in random blocks, with six treatments constituting three application techniques: conventional spraying, air-assisted spraying and air-assisted spraying combined with electrically charged droplets in two spray volumes, 50 and 100 L ha^{-1}, in four replications, totalling 24 experimental plots. The experimental plots were 24.0 x 100.0 m (width x length). The width of the plots corresponded to the boom size of the sprayer used in this research. During spraying, a self-propelling sprayer (Uniport 3000 model) was used equipped with a spray boom 24.0 m in length with hollow conical nozzles spaced every 0.35 m. The spray hollow conical nozzles used were of the JA-1 and JA-2 type, operated at working pressures of 690 and 828 kPa respectively. The spray displacement speed was 15 km h^{-1}, usually practiced by farmers in the Brazilian mid-western region (Cerrado). This sprayer operated with or without air assistance on the spray boom (conventional) combined with electric charge transference to the spray droplets in turn-on or turn-off mode. For air supply into the sleeve boom, two axial fans were positioned on the central point of the boom and operated at the maximum rotation speed. For the quantification of spray deposits, a tracer dye (Brilliant Blue) was used at a concentration of 0.3%, according to qualitative and quantitative evaluation studies of spray deposits validated by Palladini et al. (2005). The spraying of the tracer dye was performed in the R5.1 growth stage, 80 days after sowing. The average values of height and foliar area were respectively 0.92 m and 0.158 m^2 at this growth stage. The mean values of spray deposits with different application technologies on soybean plants of the Monsoy 8757 variety are shown in Tables 4 and 5. Greater spray deposits were obtained with air assistance and this spray technology combined with electrically charged droplets in relation to conventional spraying at 100 L ha^{-1} on leaflets positioned at the top part of soybean plants (Table 4). At a low volume rate, there was no observed difference in deposit levels with the different spray technologies. With the higher spray volume using air assistance combined with electric charge transference to the droplets, it was possible obtain greater spray deposits when compared to low spray volume. The best spray deposits on leaflets at the bottom position of soybean plants was obtained with air assistance combined with electric charge transference technology at a volume of 100 L ha^{-1} (Table 5). With this spray volume, air assistance technology combined with electrically charged droplets was better when compared the other two spraying technologies. There was no significant difference between the spraying techniques on tracer deposits using the volume of 50 L ha^{-1} (Table 5). These results obtained with a new spray technology, employing air assistance combined with electric charge transference to droplets, is very promising in disease management in this culture, especially for Asian soybean rust management.

Figure 7. Spraying with an air-assisted sprayer and rotating system nozzles fixed at the spray boom on soybean culture.

Table 4. Mean values of Brilliant Blue tracer deposits (μL cm^{-2}) on leaflets at the top position of soybean plants after spraying with different techniques. Botucatu, SP, Brazil, 2009/10.

Spray technique	Volume (L ha^{-1}) 100	50	LSD values ($p < 0.05$)
Air-assistance + electric charge	0.744 Aa	0.352 Ba	
Air-assistance	0.684 Aa	0.480 Aa	0.350
Conventional	0.254 Ab	0.313 Aa	
LSD values ($p < 0.05$)	0.427		
CV (%)		11.27	

Original means and data transformed in root square of x + 0.5 for analysis.
Means followed by the same letter, smaller in the column and bigger in the line, did not differ by Tukey's test at the 5% significance level.

Table 5. Mean values of Brilliant Blue tracer deposits (μL cm^{-2}) on leaflets at the bottom position of soybean plants after spraying with different techniques. Botucatu, SP, Brazil, 2009/10.

Spray technique	Volume (L ha^{-1}) 100	50	LSD values ($p < 0.05$)
Air-assistance + electric charge	1.166 Aa	0.016 Ba	
Air-assistance	0.215 Ab	0.029 Aa	0.622
Conventional	0.033 Ab	0.115 Aa	
LSD values ($p < 0.05$)	0.758		
CV (%)		21.11	

Original means and data transformed in root square of x + 0.5 for analysis.
Means followed by the same letter, smaller in the column and bigger in the line, did not differ by Tukey's test at the 5% significance level.

CONSIDERATIONS OF ASR MANAGEMENT

Despite new techniques and equipment available for the application of fungicides targeting Asian soybean rust management, other factors such as climate conditions, varieties, disease severity, plant architecture, fungicide characteristics, sowing in the same season and application time are important in ensuring culture productivity. Associated with chemical control, plant disease resistance and the adoption of the period of sowing interruption (inter-season) in most Brazilian states have contributed to decreasing the severity of Asian soybean rust. Disease monitoring time of application and choice of fungicide are important factors for the success of Asian soybean rust control.

Beforehand sowing and choosing an early variety can also contribute to the control of this disease. Nowadays, multidisciplinary research development is necessary to achieve suitable management of Asian soybean rust. Only knowledge of pesticide application techniques is not sufficient to improve control of the *P. pachyrhizi* pathogen.

REFERENCES

1. Adam, A.V. (1977). Importance of pesticide application equipment and related field practices in developing countries. In: *Pesticide management and insect resistance,* D.L. Watson & A.W. Brown, (Ed.), 217-225, Academic Press, New York
2. ANDEF – Associação Nacional de Defesa Vegetal. (2004) *Manual de tecnologia de aplicação de produtos fitossanitários.* Línea Creativa, Campinas, 50p
3. Antuniassi, U.R. (2010). Tecnologia de aplicação de defensivos: Conceitos básicos de tecnologia de aplicação de defensivos, In: *Boletim de Pesquisa de Soja 2010,* Vol.1, No.14, Fundação Mato Grosso, (Ed.), pp. 347- 372, Mato Grosso, Brazil, ISSN 1807-7676
4. Antuniassi, U.R. (2005). Tecnologia de aplicação para o controle da ferrugem da soja, *Proceedings of 1rst Workshop brasileiro sobre a ferrugem asiática,* pp. 193-219, UDUFU, Uberlândia, Minas Gerais, Brazil
5. Antuniassi, U.R.; Camargo, T.V.; Bonelli, M.A.P.O. & Romagnole, E.W.C. (2004). Avaliação da cobertura de folhas de soja em aplicações terrestres com diferentes tipos de pontas, *Proceedings of 3rd Simpósio internacional de tecnologia de aplicação de agrotóxicos,* pp. 48-51, FEPAF, Botucatu, São Paulo, Brazil
6. Bauer, F.C. (2002). *Distribuição e deposição da pulverização sob diferentes condições operacionais na cultura da soja [Glycine max (L.) Merrill].* 130p. Tese (Doutorado em Agronomia / Proteção de Plantas) – Faculdade de Ciências Agronômicas, Universidade Estadual Paulista, Botucatu, São Paulo, Brazil
7. Bauer, F.C. & Raetano, C.G. (2000). Assistência de ar na deposição e perdas de produtos fitossanitários em pulverizações na cultura da soja. *Scientia Agricola,* Vol.57, No.2, pp. 271-276, ISSN 0103-9016
8. Barros, A. L.M.B. & Menegatti, A.L. (2010). Tendências de médio prazo para o mercado de soja no Brasil e no mundo, In: *Boletim de Pesquisa*

de Soja 2010, Vol.1, No.14, Fundação Mato Grosso, (Ed.), pp. 25-27, Mato Grosso, Brazil, ISSN 1807-7676

9. Boller, W. & Raetano, C.G. (2011). *Bicos e pontas de pulverização de energia hidráulica, regulagens e calibração de pulverizadores de barras*, 31p. (Documento no prelo)
10. Brown-Rytlewski, D. & Staton, M. (2010). *Fungicide application technology for soybean rust - 2006: Field crop advisory alert*, 2p, Michigan State University Extension, Department of Plant Pathology, Michigan, USA
11. Calaça, H.A. (2007). *Ferrugem Asiática da soja: relações entre atraso do controle químico, rendimento, severidade e área foliar sadia de soja (Glycine max L. Merrill)*. 80p Dissertação (Mestrado) - Escola Superior de Agricultura Luiz de Queiroz, Piracicaba, São Paulo, Brazil
12. Christovam, R.S. (2008). *Assistência de ar e aplicação em volume baixo no controle da ferrugem asiática da soja. Botucatu*. 68p. Dissertação (Mestrado em agronomia/ Proteção de Plantas) – Faculdade de Ciências Agronômicas, Universidade Estadual Paulista, Botucatu, São Paulo, Brazil
13. Christovam, R. S.; Raetano, C.G., Aguiar-Júnior, H.O., Dal Pogetto, M.H.F.A., Prado, E.P., Junior, H.O.A., Gimenes, M.J. & Kunz, V.L. (2010a). Assistência de ar em barra de pulverização no controle da ferrugem asiática da soja. *Bragantia*, Vol.69, No.1, pp. 231-238, ISSN 0006-87065
14. Christovam, R. S.; Raetano, C.G., Dal Pogetto, M.H.F.A., Prado, E.P., Aguiar-Junior, H.O., Gimenes, M.J. & Serra, M.E. (2010b). Effect of nozzle angle and air-jet parameters in air assisted sprayer on biological effect of soybean Asian rust chemical protection. *Journal of Plant Protection Research*, Vol.50, No.3, pp. 347-353, ISSN 1427-4345
15. Christovam, R. S.; Raetano, C.G., Prado, E.P., Dal Pogetto, M.H.F.A., Aguiar-Junior, H.O., Gimenes, M.J. & Serra, M.E. (2010c). Air-assistance and low volume application to control of Asian rust on soybean crop. *Journal of Plant Protection Research*, Vol.50, No.3, pp. 354-359, ISSN 1427-4345
16. Cooke, B.K. et al. (1990). Air-assisted spraying of arable crops in relation to deposition, drift and pesticide performance. *Crop Protection*, Vol.9, No.4, pp. 303-311, ISSN 0261-2194

17. Cunha, J.P.A.R.; Reis, E.F. & Santos, R.O. (2006). Controle químico da ferrugem asiática da soja em função de ponta de pulverização e de volume de calda. *Ciência Rural,* Vol.36, No.5, pp. 1360-1366, ISSN 0103-8478
18. Hislop, E.C.; Western, N.M. & Butler, R. (1995). Experimental air-assisted spraying of a maturing cereal crop under controlled conditions. *Crop Protection,* Vol. 14, No. 1, pp. 19-26, ISSN 0261-2194
19. Jorgensen, M.K. & Witt, K.L. (1997). Spraying technique in relation to approval and use of pesticides in Northern Europe, *Proceedings of 14th Danish Plant Protection Conference: Pesticides and Environment, SP-report* No.7, 1997
20. Jorgensen, M.K. & Witt, K.L. (2000). Spraying and the impact on the environment: Spraying technique in relation to approval and use of pesticides in Northern Europe, *Proceedings of Hardi international application technology course,* Vol.1, Chap.1, pp. 4-16, Taastrup, September, 2000
21. Koch, H. (1997). The evolution of application techniques in Europe, *Proceedings of 1rst Simpósio Internacional de Tecnologia de Aplicação de Produtos Fitossanitários,* pp. 30-38, Águas de Lindóia, São Paulo, Brazil, March 26-29, 1996
22. Kunz, V.L. (2010). *Dinâmica do ar em barra pulverizadora, com saída única e dupla, deposição da calda e controle da ferrugem asiática da soja.* 48p. Tese (Doutorado em agronomia/ Proteção de plantas) - Faculdade de Ciências Agronômicas, Universidade Estadual Paulista, Botucatu, São Paulo, Brazil
23. Matthews, G.A. (1992). *Pesticide application methods* (2nd. ed.), Longman, ISBN 0-470-21818-5, London, England
24. Matthews, G.A. (2000). *Pesticide application methods,* (3rd. ed.), Blackwell Science, ISBN 9780632054732 , Oxford, UK
25. Matuo, T. (1990). *Técnicas de aplicação de defensivos agrícolas,* FUNEP, Jaboticabal, 139p.
26. Miller, P. (1997). Engineering research and development related to ground-based crop sprayers, *Proceedings of 1rst Simpósio Internacional de Tecnologia de Aplicação de Produtos Fitossanitários,* pp. 102-109, Águas de Lindóia, São Paulo, Brazil, March 26- 29, 1996

27. Monteiro, M.V.M. (2006). BVO Terrestre, In: *Manual de Operação Para Aplicações Terrestres em BVO,* 9 p., Centro Brasileiro de Bioaeronautica, Sorocaba, São Paulo, Brazil
28. Navarini, L.; Dallagnol, L.J.; Balardin, R.S.; Moreira, M.T.; Meneghetti, R.C. & Madalosso, M.G. (2007). Controle químico da ferrugem asiática (*Phakopsora pachyrhizi* Sidow) na cultura da soja. *Summa Phytopathologica,* [online], Vol.33, No.2, pp. 182-186, ISSN 0100-5405
29. Nordbo, E. (1992). Effects of nozzle size, travel speed and air assistance on artificial vertical and horizontal targets in laboratory experiments. *Crop Protection*, Vol.11, N o.3, pp. 272-277, ISSN 0261-2194
30. Ozkan H. E. (2005). Best spraying strategies to fight against Soybean Rust. Available from: http://www.jacto.com/soybean_rust.html.
31. Ozkan, H.E. (2010). Spraying Recommendations for Soybean Rust. Available at Ohio State University Extension's web site "Ohionline" (http://ohionline.usu.edu).
32. Ozkan, H.E.; Zhu, H.; Derksen, R.C.; Guler, H. & Krause, C. (2006). Evaluation of various spraying equipment for effective application of fungicides to control Asian soybean rust. *Aspects of Applied Biology*, Vol.77, pp. 1-8
33. Palladini, L.A.; Raetano, C.G. & Velini, E.D. (2005). Choice of tracers for the evaluation of spray deposits. *Scientia Agricola*, Vol.62, No.5, pp. 440-445, ISSN 0103-9016
34. Prado, E.P. 2009. *Assistência de ar em pulverização no manejo fitossanitário na cultura da soja [Glycine max (L.) Merrill.]*. 94p. Dissertação (Mestrado em agronomia/ Proteção de Plantas) – Faculdade de Ciências Agronômicas, Universidade Estadual Paulista. Botucatu, São Paulo, Brazil
35. Prado, E. P.; Raetano, C.G.; Aguiar-Junior, H.O.; Dal Pogetto, M.H.F.A.; Christovam, R.S.; Gimenes, M.J. & Araújo, D. (2010). Velocidade do ar em barra de pulverização na deposição da calda fungicida, severidade da ferrugem asiática e produtividade da soja. *Summa Phytopathologica*, Vol.36, No.1, pp. 45-50, ISSN 0100-5405
36. Prado, E.P. (2011). Different fungicides application technologies to control of Asian rust in the soybean culture, 120p. (Not published)
37. Raetano, C.G. (2002). Assistência de ar em pulverizadores de barra. *O Biológico*, Vol.64, No.2, pp. 221-225, ISSN 0366-0567

38. Raetano, C.G. & Bauer, F.C. (2003). Efeito da velocidade do ar em barra de pulverização na deposição de produtos fitossanitários em feijoeiro. *Bragantia*, Vol.62, No.2, pp. 329- 334, ISSN 0006-8705
39. Raetano, C.G. & Bauer, F.C. (2004). Deposição e perdas da calda em feijoeiro em aplicação com assistência de ar na barra pulverizadora. *Bragantia*, Vol.63, No.2, pp.309-315, ISSN 0006-8705
40. Raetano C.G. & Merlin A. (2006). Avanços tecnológicos no controle da ferrugem da soja, In: *Ferrugem Asiática da Soja*, L. Zambolim, (Ed.), 115–138, UFV, ISBN 85-60027-14-9, Viçosa, Minas Gerais, Brazil
41. Robinson, T.H. (1993). Large-scale ground-based application techniques, In: *Application technology for crop protection*, G.A. Matthews & E.C. Hislop, (Eds.), 163-186, CAB International, ISBN 0-85198-834-2, Wallingford, Oxon, UK
42. Sartori, S. (1997). Equipamentos tratorizados para culturas de baixo fuste: situação no Cone– Sul, *Proceedings of 1rst Simpósio Internacional de Tecnologia de Aplicação de Produtos Fitossanitários*, pp. 110-112, Águas de Lindóia, São Paulo, Brazil, March 26-29, 1996
43. SCudeler, F. & Raetano, C.G. (2004). Assistência de ar e angulação da barra pulverizadora na deposição e perdas da pulverização na cultura da batata. Depto. de Produção Vegetal, Faculdade de Ciências Agronômicas, Universidade Estadual Paulista, São Paulo, Brazil, 35p. (Relatório Científico).
44. Silva, M. A. S. (2001). *Depósitos da calda de pulverização no solo e em plantas de tiririca (Cyperus rotundus L.) em diferentes condições de aplicação*. 53p. Tese (Doutorado em Agronomia / Agricultura) – Faculdade de Ciências Agronômicas, Universidade Estadual Paulista, Botucatu, São Paulo, Brazil
45. Soares, R.M.; Rubin, S.A.L.; Wielewicki, A.P. & Ozelame, J.G. (2004). Fungicidas no controle da ferrugem asiática (*Phakopsora pachyrhizi*) e produtividade da soja. *Ciência Rural*, Vol.34, No.4, pp.1245-1247, ISSN 0103-8478
46. Taylor, W.A. & Andersen, P.G. (1991). Enhancing conventional hydraulic nozzle use with the Twin Spray System. *British Crop Protection Council Monograph*, No.46, pp. 125- 136, UK
47. Taylor, W.A. & Andersen, P.G. (1997). A review of benefits of air assisted spraying trials in arable crops. *Aspects of Applied Biology*, Vol.48, (January 1997), pp. 163-174, ISSN 0265-1491, Wellesbourne, Warwick, UK

48. Taylor, W.A.; Andersen, P.G. & Cooper, S. (1989). The use of air assistance in a field crop sprayer to reduce drift and modify drop trajectories, *Proceedings of 3rd Brighton Crop Protection Conference Weeds*, p.631, ISBN 0948404361, British Crop Protection Council, Brighton, Farnham, UK, November 20-23, 1989
49. Tomazela, M.S.; Martins, D.; Marchi, S.R. & Negrisoli, E. (2006). Avaliação da deposição da calda de pulverização em função da densidade populacional de *Brachiaria plantaginea*, do volume e do ângulo de aplicação. *Planta Daninha*, Vol.24, No.1, pp.183-189, ISSN 1806-9681
50. Van De Zande; J.C.; Meier, R. & Van Ijzendoorn, M.T. (1994). Air-assisted spraying in winter wheat-results of deposition measurements and the biological effect of fungicides against leaf and ear diseases. *Proceedings of British Crop Protection Conference - pests and diseases*, pp.313-318, BCPC, Brighton, UK, 1994
51. Zhu, H.; Brazee, R.D.; Fox, R.D.; Derksen, R.C. & Ozkan, H.E. (2008a). Development of a canopy opener to improve spray deposition and coverage inside soybean canopies: Part 1. Mathematical models to assist opener development. *Transactions of the ASABE*, Vol.51, No.6, pp.1905-1912, ISSN 2151-0032
52. Zhu, H.; Derksen, R.C.; Ozkan, H.E.; Reding, M.E. & Krause, C.R. (2008b). Development of a canopy opener to improve spray deposition and coverage inside soybean canopies: Part 2. Opener design with field experiments. *Transactions of the ASABE*, Vol.51, No.6, pp.1913-1921, ISSN 2151-0032
53. Yorinori, J.T.; Júnior, J.N. & Lazzarotto, J.J. (2004). Ferrugem asiática da soja no Brasil: Evolução, importância econômica e controle, 36p, EMBRAPA Soja, Londrina, Paraná, Brazil (Documentos, 247)
54. Yorinori J.T. (2005). A ferrugem asiática da soja no continente americano: evolução, importância econômica e estratégias de controle. *Proceedings of Workshop Brasileiro sobre a Ferrugem Asiática*, pp.21-37, EDUFU, Uberlândia, Minas Gerais, Brazil, 2005
55. Yorinori, J.T.; Yuyama, M.M. & Siqueri, F.V. (2010). Doenças da soja, In: *Boletim de Pesquisa de Soja 2010*, Vol.1, No.14, Fundação Mato Grosso, (Ed.), pp. 218-274, Mato Grosso, Brazil, ISSN 1807-7676
56. Yorinori, J.T. (2010). Controle da ferrugem "Asiática" da soja na safra 2006/2007, 3p, EMBRAPA soja, Londrina, Paraná, Brazil.

5

CHAPTER

INCREASED LEVELS OF ANTINUTRITIONAL AND/OR DEFENSE PROTEINS REDUCED THE PROTEIN QUALITY OF A DISEASE-RESISTANT SOYBEAN CULTIVAR.

Daniele O. B. Sousa [1,†], Ana F. U. Carvalho [2,*], José Tadeu A. Oliveira [1,†], Davi F. Farias [1,†], Ivan Castelar [3,†], Henrique P. Oliveira [1,†] and Ilka M. Vasconcelos

[1]*Department of Biochemistry and Molecular Biology, Federal University of Ceará, Campus do Pici, Fortaleza 604440-900, CE, Brazil*

[2]*Department of Biology, Federal University of Ceará, Campus do Pici, Fortaleza 60440-900, CE, Brazil*

[3]*Department of Finance, Federal University of Ceará, Campus Benfica, 60440-970, Fortaleza 60020-180, CE, Brazil*

ABSTRACT

The biochemical and nutritional attributes of two soybean (Glycine max (L.) Merr.) cultivars, one susceptible (Seridó) and the other resistant (Seridó-RCH) to stem canker, were examined to assess whether the resistance to pathogens was related to levels of antinutritional and/or defense proteins in the plant and subsequently affected the nutritional quality. Lectin, urease, trypsin inhibitor, peroxidase and chitinase activities were higher in the resistant cultivar. Growing rats were fed with isocaloric and isoproteic diets prepared with defatted raw soybean meals. Those on the Seridó-RCH diet showed the worst performance in terms of protein quality indicators. Based on regression analysis, lectin, trypsin inhibitor, peroxidase and chitinase appear to be involved in the resistance trait but also in the poorer nutritional quality of Seridó-RCH. Thus, the development of cultivars for disease resistance may lead to higher concentrations of antinutritional compounds, affecting the quality of soybean seeds. Further research that includes the assessment of more cultivars/genotypes is needed.

Keywords: Seridó-RCH; unintended traits; conventional breeding; nutritional value; plant protein

1. INTRODUCTION

Natural genetic variation within species and among related species has been one of the main sources of diversity for crop improvement [1]. For centuries, conventional plant breeding programs have produced new traits, higher yields and improved quality. However, little attention has been paid to metabolic changes occurring in successive generations. This issue has gained importance only recently in the context of safety assessment of genetically modified (GM) crops [2]. This issue is also of great significance because several studies have reported wide variation in the chemical and nutritional composition of genetically close cultivars of important legume seeds, such as soybeans, obtained by conventional plant breeding [3,4]. Nevertheless, little is known about the consequences of the observed differences resulting from plant breeding and the impact of the possible unintended effects on the nutritional value of crops.

The impairment of legumes' nutritional value is attributed in part to the presence of different compounds classically known as toxic and/or antinutritional factors, which act as direct or indirect antagonists of nutrient availability [5]. For soybean seeds, trypsin inhibitor (SBTI) and lectin (SBA) are considered to be the major proteins responsible for poor nutritional value [6]. In fact, soybean varieties deficient in Kunitz protease inhibitor or in lectin have shown improved nutritional quality [7,8]. It has been suggested that other proteins, such as soybean toxins that are lethal to mice when administered by the intraperitoneal route and urease, might also contribute to the deleterious effects observed upon feeding soybeans to rats [4]. Of lesser significance are antinutritional factors that are non-protein in nature, such as goitrogens, tannins, phytoestrogens, flatus-producing oligosaccharides, phytate, and saponins [9]. In the present work, we focused only on protein antinutritional factors.

The defensive role of lectins and trypsin inhibitors against phytopathogens and insects is well known [10,11], but in contrast to the toxic/antinutritional effects, this role cannot be effective when the factors are present at very low levels, which in turn could depress plant performance in general. In addition, reduced levels can decrease plant performance in the field. Considering the scarcity of information on the correlation of plant pathogen resistance and nutritional value, there is a need for studies to ascertain whether the proteins possibly responsible for resistance interfere with the nutritional quality of the cultivar.

In the above context, the soybean cultivar Seridó-RCH was obtained through conventional plant breeding (backcrossing of the Seridó cultivar) by EMBRAPA (Brazilian Agricultural Research Corporation) in experimental fields located in Balsas, Maranhão, Brazil. The parental cultivar Seridó (BR 28) corresponds to the progeny-F6, identified by the name BR 83–9221, which originated from the crossing of Santa Rosa with BR 78–11202. The resulting seeds in F1 were harvested and used to produce F2 and successive crosses of segregating generations through planting and harvesting to the desired level of homozygosis, which was achieved by F6 inbreeding. Seridó (BR 28) possesses the resistance trait against frogeye leaf spot caused by Cercospora sojina, but it is susceptible to both stem canker and bacterial pustules caused by Diaporthe phaseolorum f.sp. meridionalis and Xanthomonas axonopodis pv. glycines, respectively. Seridó-RCH, also denominated BR 96–4909 or MA/BRS-165 cultivar, was obtained by backcrossing of BR 28 (Seridó) with the Embrapa 20 (Doko RC) cultivar for incorporation of resistance to stem canker. In fact, Seridó-RCH was shown to be resistant to stem canker and frogeye leaf spot and also to bacterial pustules. These are three major diseases for soybean culture [12]. Nevertheless, nothing is known about the composition or nutritional value of these cultivars and whether the acquired high resistance to pathogens is associated with certain unintended nutritional effects that could have emerged as a result of the process of backcrossing to obtain Seridó-RCH. This is possible for antinutritional factors, as a similar outcome was observed following the emergence of resistance to bacterial pustules when backcrossing was performed to obtain Seridó-RCH. Thus, this work aimed to perform a comparative study with emphasis on the protein quality of both Seridó and Seridó-RCH.

Different markers and approaches can be used to assess the nutritional quality of proteins, such as the Protein Digestibility-Corrected Amino Acid Score (PDCAAS) and the indispensable Amino Acid Score (AAS) [13]. However, animal feeding studies may provide additional useful information to complement safety and nutritional value assessments of food and feed, especially when unintended effects are suspected. Additionally, in vivo assays allow the analysis of crucial parameters related to protein quality, such as the balance between the amount of nitrogen retained and lost by animal tissues. Moreover, it is possible to observe systemic effects that can culminate with poor development or that can cause alterations in the vital organs of animals. Thus, in the current study, a traditional short-term feeding trial was organized in which we could measure several parameters, such as food intake, body weight gain, net protein utilization (NPU), protein digestibility and biological value. The rat was selected for the study because this species has been traditionally used to assess the safety of food. In addition, the levels of several classical antinutritional and defense proteins were measured, and the results were correlated to the nutritional responses of the feeding trial.

2. EXPERIMENTAL SECTION

2.1. Biological and Chemical Reagents

Soybean seeds of the Seridó and Seridó-RCH cultivars were developed and supplied by EMBRAPA (Experimental Field at Balsas, Maranhão, Brazil). The sowing was performed in soil (Oxisol) previously fertilized with NPK (2:20:20) at a dose of 400 kg/ha. Seeds (5 kg per cultivar) were obtained from three different plots. The plots consisted of 8 rows, each 8 m in length and equally spaced at 0.5 m apart, with a population density of 250,000 plants/ha maintained under continuous irrigation from planting until harvesting. Seeds were sowed in mid-November and harvested from March to April 2010, when the plants reached the R8 development stage and moisture content between 14% and 16%. After harvesting, the seeds were stored in sealed packages in a ventilated area free of fungi and rodents, and the temperature and humidity were maintained below 25 °C and 70%, respectively.

Male Wistar rats were obtained from outbred colonies maintained at the Federal University of Ceará (Fortaleza, Brazil). This institution has adopted Poiley's system of breeding to avoid consanguinity and to preserve colony heterogeneity. The animals were maintained in an appropriate climatized environment, with controlled conditions of temperature (22 ± 2 °C), humidity (55%) and photoperiod (12 h of light/12 h of dark). Prior to the experiments, the animals were maintained with a pelletized diet and drinking water, both of which were sterilized. Rabbit blood was obtained from animals in colonies maintained at the same institution.

Egg whites (EWs) from chicken (Sigma E0500), β-glucuronidase (Sigma G0251), guaiacol (Sigma G5502), laminarin (Sigma L9634), l-BAPNA (Sigma B3133), N-acetyl-d-glucosamine (Sigma A8625), soybean trypsin inhibitor (Sigma T9128), trypsin (Sigma T6424) and urease (Sigma U0251) were purchased from Sigma-Aldrich Co. LLC (St. Louis, MO, USA). All other chemical reagents used in the experiments were of analytical grade.

2.2. Determination of Classical Antinutritional Proteins

Mature seeds were ground in a coffee grinder fitted with a 1-mm mesh screen, and the resulting flour was treated with petroleum ether (1:10, w/v) to extract lipids that could interfere with the protein determination. The solvent petroleum ether was evaporated to dryness under a gentle stream of air at room temperature (22 ± 3 °C). The defatted flour was extracted with 0.025 M Tris-HCl, pH 7.5 (1:5, w/v) [14]. The crude extract obtained was used for detection of the lectin trypsin inhibitor and urease activities. Lectin activity was assessed following a serial two-fold dilution of the samples [14]. The extracts were diluted with 0.15 M NaCl and mixed with a 2% suspension of rabbit erythrocytes prepared in 0.15 M NaCl. The degree of agglutination was monitored visually after the tubes had been left to stand at 37 °C for 30 min and after an additional 30 min at room temperature (22 ± 3 °C). The minimal protein concentration after serial dilution that still promoted agglutination visible to the naked eye was used to calculate the lectin equivalents. The lectin activity was expressed as grams of lectin equivalents per kilogram of soybean flour. Trypsin inhibitor activity was examined using trypsin and l-BAPNA as substrate. The activity was expressed as the amount (g) of trypsin inhibited per kilogram of soybean flour [15]. Urease assay was assessed using the procedure described by Kaplan, with minor modifications, and the urease activity was expressed as units of enzyme per kilogram of soybean flour [16]. For preparation of urease solution (50 μg/mL), glycerol was added to 0.2 M phosphate buffer, pH 6.5, containing 2% EDTA, at a ratio of 1:4 to preserve this solution for about three weeks at 4 °C. In addition, to obtain a standard curve, urea was used at 0.5 M instead of 0.3 M and prepared in the same buffer above.

2.3. Determination of Classical Plant Defense Proteins

Crude extracts (1:5, w/v) were prepared with 0.025 M Tris-HCl, pH 7.5, as described below and were used for the detection of chitinase, peroxidase and β-1,3-glucanase. The determination of peroxidase activity was performed using 0.02 M guaiacol and 0.06 M hydrogen peroxide as a donor and an acceptor of protons, respectively. An increase of one absorbance unit per minute at 480 nm was used as a unit of peroxidase activity (1 UAP) [17]. Chitinase activity determination in the crude extract was carried out by measuring the N-acetyl-d-glucosamine released due to hydrolytic enzyme action on colloidal chitin. Chitinase activity was expressed as nanokatal (nkat) per gram of seed flour (nkat/g) and was defined as the amount of enzyme that releases 1 nmol N-acetyl-d-glucosamine per second

[18]. β-1,3-Glucanase activity was expressed in nkat per gram of seed flour (nkat/g) and was defined as the amount of enzyme necessary to catalyze the formation of 1 nmol of glucose equivalent per second [18].

2.4. Amino Acid Composition

Defatted soybean flour was hydrolyzed with 6 M HCl containing 10 g/L phenol at 110 °C for 22 h in sealed glass tubes under a N_2 atmosphere. HCl and phenol were removed by evaporation, and the amino acid composition was established after chromatography on a Biochrom 20 system (Pharmacia). Tryptophan was determined according to the method described by Pinter-Szakács and Molnár-Perl [19]. All determinations were run in triplicate.

2.5. Diets

All soybean diets were prepared using seeds produced under strict quality standards. In addition, as the main concern of this work was to compare the nutritional quality of the seed proteins of two soybean cultivars, any type of processing (such as autoclaving or irradiation) that could cause changes in compounds, including anti-nutritive proteins, was not applied, except for the process of defatting the soybean flours. This step was performed to allow formulation of the 10% protein test diets with the same amount and source of lipids (maize oil) as in the control diet. Aflatoxin-free soybean (Seridó and Seridó-RCH) samples, confirmed by analysis carried out with a commercial competitive enzyme-linked immunosorbent assay (ELISA) kit [20], were ground in a hygienized blender and then refined using a coffee grinder (mesh size 1.0 mm). Powdered diets (1 kg) were formulated to contain the equivalent of 100 g protein per kilogram of diet (Table 1) in the form of EW protein or seed meal from cv. Seridó-RCH or Seridó. Diets containing the respective seed flours were supplemented with l-tryptophan and l-methionine based on the amino acid contents of the raw seeds to bring the amino acid content to the target requirements for rats [21]. Diet ingredients were mixed exhaustively in a nutrition laboratory in a flour homogenizer (SY-SM10 Spiral Mixer, Sunrry Machinery Technology Co., Ltd., Guangzhou, China) under adequate sanitary conditions and then analyzed for protein content to confirm the theoretical values. Before starting the feeding trial, the animals were submitted to a 5-day period of adaptation to the powdered diets. This protocol had been previously established in studies performed in metabolic cages suited for powdered diets (Nalgene Metabolic Cages, Buguggiate, Italy). A diet containing no protein (non-protein control, or NPC) was fed to a group of animals to allow determination of certain nutritional parameters.

Table 1. Composition (g/kg) of non-protein control (NPC), egg white (EW) and experimental diets.

Ingredient	NPC	EW	Seridó	Seridó-RCH
maize starch	555.6	422.4	409.4	399.6
cassava starch	111.1	111.1	100.0	100.0
glucose	166.7	166.7	150.0	150.0
maize oil	55.6	55.6	50.0	50.0
vitamin mix [a]	55.6	55.6	50.0	50.0
mineral mix [b]	55.6	55.6	50.0	50.0
EW	-	133.1	-	-
Seridó	-		187.6	-
Seridó-RCH	-		-	197.4
L-methionine [c]	-		2.3	2.3
L-tryptophan [c]	-		0.7	0.7
Energy [d] (kcal/g)	3.4	3.4	3.6	3.6

[a] Vitamin mix (g/kg diet): vitamin B_{12} (100%), 0.02; folic acid, 0.04; biotin (1%), 4.0; pyridoxine HCl, 0.04 thiamine HCl, 0.06; riboflavin (99%), 0.21; Ca-pantothenate (45%), 1.2; nicotinic acid, 4.0; inositol, 4.0; *p*-amino-benzoic acid, 12.0; choline chloride (50%), 24.0; maize starch, 950.43. [b] Mineral mix (g/kg diet): calcium citrate, 296.1; calcium carbonate (40%), 65.8; copper carbonate, 1.1; magnesium carbonate, 34.3; zinc carbonate, 0.48; ferric citrate, 9.1; magnesium chloride.$6H_2O$, 5.82; sodium chloride, 74.0; potassium chloride, 119.5; monobasic calcium phosphate, 108.2; dibasic potassium phosphate, 210.1; sodium fluoride, 0.48; potassium iodate 0.1; magnesium sulfate, 75.4. [c] Diets containing raw seed flour of the distinct cultivars were supplemented with L-methionine and L-tryptophan according to their amino acid compositions. [d] The energy content was estimated by multiplying the percentages of crude protein, crude fat, and carbohydrates by their respective modified Atwater factors.

2.6. Feeding Trials

Traditionally, animal feeding trials focused on protein quality are performed in growing male rats. In those trials, the parameters used are daily food intake, body weight gain, relative organ weights, net protein utilization (NPU), apparent protein digestibility and biological value. These are well-established parameters based on previous studies in which no difference in the responses of male and female animals was significant [22]. Therefore, in the present study, male Wistar rats were weaned at 21 days of age and given a commercial stock diet until their weights reached 55–60 g. Next, they were fed with the EW diet ad libitum for 3 days as a period of adaptation to pulverized diets and were then selected according to food consumption and body weight. The animals were divided into 4 groups of twelve rats each, housed individually in screen-bottomed cages and fed with control (EW), non-protein (NPC) or experimental diets (Seridó-RCH or Seridó) for 10 days. Feed and water were supplied ad libitum under adequate sanitary conditions. The rats' weights, diet spillage and refused diets were recorded daily. Feces were collected during the last 5 days of the experimental period, bulked, freeze dried, weighed and ground in a coffee grinder. At the end of the trial, the rats were euthanized by halothane overdose, and their internal organs were dissected. The contents of the stomachs and intestines were rinsed three times with 0.01 M phosphate buffer solution (PBS) containing 0.15 M NaCl, pH 7.4. The organs were then freeze dried, whereas the carcasses were dried in an oven at 100 °C for 24 h. Dry weights were recorded before incorporating the organs with their original carcasses, which were then ground and kept in a desiccator for appropriate analyses. This study was approved by the Animal Experimentation Ethics Committee of Federal University of Ceará (CEPA) in accordance with the N.11.794/2008 Act, which governs the breeding and use of laboratory animals for teaching and research purposes across the country.

2.7. Chemical Analyses

Diets, carcasses and ground fecal samples were analyzed for moisture content and total nitrogen [23,24]. The data were used to calculate apparent protein digestibility (D), NPU and biological value (BV) [22]. The results were calculated for each rat, and the mean within each group was determined.

2.8. Statistical Analyses

The results were subjected to a one-way analysis of variance. The significance of differences between two means was determined by Student's t test, and the Tukey honest test was used when comparing multiple means. Multiple regression analysis was applied, relating trypsin inhibitor, lectin, peroxidase, chitinase, β-1,3-glucanase and urease to food intake, weight gain, NPU and digestibility. Due to the high degree of multicollinearity presented in the samples among the independent variables (lectin, trypsin inhibitor, urease, peroxidase, chitinase and β-1,3-glucanase), a technique of orthogonalization was employed [25]. Principal components were constructed, and a special test was performed to discard non-significant components [26].

3. RESULTS

3.1. Antinutritional and/or Defense Proteins

The activities related to several classical antinutritional and/or defense proteins in the soybean cultivars are presented in Table 2. Trypsin inhibitor activity (g of trypsin inhibited per kg of flour) was significantly higher in the crude extract from the Seridó-RCH cultivar than in that from Seridó. The highest hemagglutinating activity (g of lectin equivalents per kg of flour) measured against rabbit erythrocytes was detected for Seridó-RCH compared with the Seridó cultivar. Regarding urease activity (units of enzyme per kg flour), it was observed that the value for Seridó-RCH was approximately five-fold that for Seridó. Table 2 also shows the activities of proteins related to pathogenesis (PR proteins), such as peroxidase and chitinase. The results showed that seeds from the resistant cultivar Seridó-RCH had higher levels of constitutive peroxidase and chitinase activities compared with seeds from Seridó. The crude extracts of Seridó-RCH and Seridó did not show β-1,3-glucanase activity.

3.2. Amino Acid Composition

For all the studied amino acids, no significant difference was observed between Seridó and Seridó-RCH (Table 3). Except for tryptophan, both cultivars contain adequate levels of essential amino acids when compared with the requirements for children [27]. In comparison with requirements for rats, except for lysine and phenylalanine + tyrosine, both cultivars were deficient in all essential amino acids, with tryptophan and methionine + cysteine being the first and second limiting amino acids, respectively. Thus, the diets containing the raw seed flours were supplemented with tryptophan and methionine to eliminate the deficiency of these constituents.

3.3. Nutritional Parameters

Although daily food intake was not significantly different between the test diets and the EW control, the nutritional parameters were significantly lower for Seridó and Seridó-RCH compared with those calculated for the EW control. The body weight gain of the rats fed with test diets were compared with those of animals fed with EW and NPC diets (Table 4). Both test diets caused body weight gain much lower than that observed for the EW control but were able to promote animal growth, in contrast to the NPC diet. Obviously, this was an expected result because we were comparing a reference animal protein with plant proteins, and specifically legume seed proteins [28]. The Seridó diet presented higher values of NPU (48.2%), protein digestibility (63.8%) and body nitrogen (8.5 g/kg body weight) compared with Seridó-RCH, which showed values of 28.8%, 52.7% and 6.8 g/kg, respectively.

Table 2. Biological activities present in the crude extracts of soybean cultivars *.

Activity	Cultivar	
	Seridó	Seridó-RCH
trypsin inhibitor [a]	37.50 ± 1.10 a	59.71 ± 1.40 b
Lectin [b]	0.10 ± 0.01 a	0.45 ± 0.01 b
Urease [c]	132,000 ± 11.32 a	630,000 ± 17.44 b
Peroxidase [d]	0.29 ± 0.04 a	3.92 ± 0.01 b
Chitinase [e]	17.76 ± 0.90 a	23.50 ± 2.00 b
β-1,3-glucanase [f]	ND [g]	ND

* The values are the mean ± standard deviation of triplicate samples. Values with different letters in the same row differ significantly ($p < 0.05$). [a] Trypsin inhibitor activity is expressed as g of trypsin inhibited per kg of flour. [b] Lectin activity is expressed as g of lectin equivalents per kg of flour. [c] Urease activity is shown as units of enzyme per kg of flour. The units were calculated from information that 1 g of pure enzyme contains 870,000 units, as indicated by Sigma. [d] One unit of peroxidase activity was defined as the amount of enzyme that causes an increase of one absorbance unit per minute at 480 nm. [e] Chitinase activity was expressed as nanokatal (nkat) per gram of flour (nkat/g) and was defined as the amount of enzyme that releases 1 nmol *N*-acetyl-D-glucosamine per second. [f] β-1,3-glucanase activity was calculated as nkat per gram of flour (nkat/g) and was defined as the amount of enzyme catalyzing the formation of 1.0 nm glucose equivalent per second. [g] Not detected.

The relative organ weights (g per 100 g of body weight) are presented in Table 5. The diets based on Seridó and Seridó-RCH led to organ weight changes in comparison with the internal organs of EW-fed rats. In general, the experimental diets caused atrophy of the thymus and spleen and enlargement of the small intestine and pancreas based on alterations of their relative dry weights. These changes were even more significant in rats fed on a Seridó-RCH diet. The rats fed on an NPC diet showed divergence in the weights of various organs in relation to the rats fed on EW diet, which can be explained by protein starvation.

Table 3. Comparison of the amino acid composition (g/kg of protein) of the soybean cultivars with WHO/FAO/UNU [27] patterns of amino acid requirements for children and with those required for rats [21] *.

Amino acid	Soybeans		Rat requirement	Requirement for children (3–10 years)
	Seridó	Seridó-RCH		
Essential				
Thr	37.0	38.5	40	25
Val	46.8	47.2	55	40
Ile	37.7	39.8	50	31
Leu	76.4	76.1	80	61
Lys	65.8	68.4	60	48
Phe + Tyr	107.2	107.0	90	41
Met + Cys	30.3	30.3	45	24
Trp	7.5	8.8	15	6.6
Non-essential				
Asx	116.7	118.8		
Glx	188.4	186.4		
Ser	42.5	41.5		
Gly	38.7	39.3		
Ala	41.2	41.9		
His	29.0	30.3	25	16
Arg	81.5	74.7	50	
Pro	53.1	51.0		

* The values are means of a duplicate. SD values were omitted because they were less than 5% of the means.

Table 4. Nutritional parameters of rats fed on Seridó and Seridó-RCH seed-based defatted flours compared with those of rats fed on egg white (EW) and non-protein control (NPC) diets *.

	Diets			
	NPC	EW	Seridó	Seridó-RCH
initial body weight (g)	71.7 ± 2.8 a	71.8 ± 2.6 a	71.3 ± 2.2 a	71.7 ± 2.4 a
final body weight (g)	56.6 ± 2.9 c	113.4 ± 8.5 a	85.9 ± 7.0 b	81.4 ± 6.4 b
body weight gain (g)	−15.2 ± 1.5 d	42.6 ± 1.7 a	14.4 ± 0.8 b	9.8 ± 1.2 c
daily food intake (g)	2.1 ± 0.6 b	3.7 ± 0.6 a	3.5 ± 0.7 a	3.7 ± 0.7 a
net protein utilization (%)	-	90.7 ± 1.2 a	48.2 ± 1.0 b	28.8 ± 1.0 c
protein digestibility (%)	-	95.4 ± 1.0 a	63.8 ± 1.0 b	52.7 ± 1.6 c
biological value (%)	-	95.1 ± 2.3 a	75.6 ± 0.4 b	54.7± 3.5 c
body nitrogen (g/kg)	8.0 ± 0.8 c	10.3 ± 1.0 a	8.5 ± 0.4 b	6.8 ± 0.4 d

* The values are the mean ± standard deviation ($n = 12$). Values with different letters in the same row differ significantly ($p < 0.05$).

Table 5. Relative organ dry weights (g/100 g body dry matter) of rats fed with non-protein control (NPC), egg white (EW) and experimental diets *.

Organ	Diets			
	NPC	EW	Seridó	Seridó-RCH
stomach	0.67 ± 0.03 b	0.61 ± 0.04 ac	0.63 ± 0.02 ab	0.58 ± 0.04 c
intestine	2.42 ± 0.44 a	2.25 ± 0.20 a	2.83 ± 0.22 b	3.24 ± 0.13 c
cecum + colon	0.67 ± 0.04 b	0.70 ± 0.04 ab	0.74 ± 0.06 a	0.74 ± 0.06 a
liver	4.52 ± 0.35 b	5.19 ± 0.44 a	4.29 ± 0.25 b	4.57 ± 0.32 b
pancreas	0.28 ± 0.02 a	0.27 ± 0.02 a	0.40 ± 0.04 b	0.51 ± 0.05 c
thymus	0.14 ± 0.01 d	0.28 ± 0.02 a	0.23 ± 0.02 b	0.19 ± 0.01 c
spleen	0.14 ± 0.01 b	0.25 ± 0.01 a	0.16 ± 0.01 b	0.13 ± 0.02 b
kidneys	0.90 ± 0.05 b	0.79 ± 0.06 a	0.75 ± 0.04 a	0.77 ± 0.04 a
heart	0.37 ± 0.04 b	0.29 ± 0.01 a	0.29 ± 0.02 a	0.31 ± 0.02 a
lungs	0.49 ± 0.05 a	0.47 ± 0.03 a	0.46 ± 0.05 a	0.45 ± 0.02 a

* The values are the mean ± standard deviation ($n = 12$). Values with different letters in the same row differ significantly ($p < 0.05$).

3.4. Relationships between Unintended Effects and Nutritional Performance

The results of the regression analyses are depicted in Table 6. The technique of orthogonalization used in this study to correct for the high degree of multicollinearity presented by the data only works when collinearity is not perfect. This was not the case for the variable urease. After discarding this variable, principal components were constructed from lectin, trypsin inhibitor, peroxidase and chitinase, and regressions of the nutritional parameters were run on the principal components retained. From these regressions, the coefficients of the original independent variables were recovered. This analysis showed that besides lectin and trypsin inhibitor, peroxidase and chitinase were also significant explanatory variables for the protein quality indicators. In the two cultivars, all of these proteins significantly influenced the food intake in the reverse direction at the same magnitude (−0.60, −0.61, −0.59 and −0.60 were the respective estimated coefficients). Lectin, trypsin inhibitor and chitinase had a significant negative influence on body weight gain, but the effect of the first two proteins was much higher than that of chitinase, as shown by their respective estimated coefficients (−3.33, −3.01 and −1.96, respectively). Regarding NPU, lectin, trypsin inhibitor and chitinase had a negative influence, with estimated coefficients of −2.55, −0.62 and −4.81, respectively. Additionally, lectin and trypsin inhibitor had a negative effect on digestibility (estimated coefficients of −8.43 and −5.22, respectively), whereas chitinase had a positive impact (estimated coefficient of 3.99).

Peroxidase did not show any significant effect on body weight gain, NPU or digestibility.

However, in the case of organ weights, the variables urease, lectin, trypsin inhibitor, peroxidase and chitinase presented a degree of collinearity, which prevented any regression analysis, even with orthogonalization techniques. Therefore, regression analysis is not presented to explain the organ weights.

Table 6. Regression results: estimated coefficients, T-ratios and R^2.

Independent variable	Dependent variables			
	Diet Intake	Weight Gain	NPU	Digestibility
lectin	−0.60 (−6.45) *	−3.33 (−4.27) *	−2.55 (−36.76) *	−8.43 (−8.26) *
trypsin inhibitor	−0.61 (−6.45) *	−3.01 (−4.63) *	−0.62 (−18.25) *	−5.22 (−10.41) *
peroxidase	−0.59 (−6.45) *	2.86 (1.69)	−0.04 (−0.48)	0.14 (0.13)
chitinase	−0.60 (−6.45) *	−1.96 (−7.79) *	−4.81 (−47.88) *	3.99 (2.7) *
R^2	0.81	0.93	0.99	0.98

* Significant at $\alpha = 0.05$.

4. DISCUSSION

The results showed that the resistant cultivar obtained by conventional breeding did not have its amino acid profile altered, although there were changes in the activities of certain antinutritional and/or defense-related proteins. Indeed, the activities of all the analyzed proteins were higher in the more resistant (Seridó-RCH) soybean cultivar, except for β-1,3-glucanase, whose activity was not detected in either cultivar. The trypsin inhibitor content in the seed flour of Seridó-RCH was nearly 63% higher than that in the Seridó cultivar. Trypsin inhibitors are proteins widely distributed in plant species. Their roles are diverse, including one of the most important defense strategies used by plants to combat phytophagous insects and microorganisms [29,30]. It is possible that the trypsin inhibitor of soybean belongs to the arsenal of molecules related to the higher resistance of Seridó-RCH against fungi. The action of certain protease inhibitors against bacteria has also been reported [31]. Nevertheless, to the knowledge of the authors, no association of soybean trypsin inhibitors with activity against bacteria has been described so far. Regarding lectins, they belong to a class of protein also related to plant defense against insects, fungi, bacteria, and nematodes [32,33]. Although many plant lectins are toxic to phytopathogenic fungi and bacteria, there are no reports specifically on soybean lectins with these activities. Therefore, the resistance of soybean to these organisms is apparently not related to this class of protein. However, the Seridó-RCH cultivar presented higher levels of lectin compared with the Seridó cultivar in the present study. This is perhaps an example of unintended effects obtained after the process of conventional breeding.

Ureases are metalloenzymes present in virtually all plant species and ubiquitously distributed in plant tissues [34]. An embryo-specific soybean urease has been shown to impair growth of selected phytopathogenic fungi at sub-micromolar concentrations [35]. Moreover, it has been shown that soybean plants with no urease were more susceptible to fungi than plants containing normal levels of the enzyme [36]. Thus, our data indicate that urease may be involved in the resistance of Seridó-RCH to fungi.

Regarding peroxidase, chitinase and β-1,3-glucanase, these proteins are related to defense (PR proteins) and can be present constitutively in plants and are therefore involved in resistance to herbivores and pathogens [37]. The levels of these proteins were analyzed in the seed flour of both cultivars in the current study. Peroxidase, for example, presented a specific activity 14 times higher in the Seridó-RCH cultivar than in the Seridó cultivar. This enzyme is involved in plant defense against several stress agents, and its level can be associated with higher or lesser plant resistance [38]. There are also reports of peroxidase with direct toxicity against certain species of fungi and bacteria [39]. Seridó-RCH also presented higher chitinase activity compared with Seridó. The natural substrate of chitinases, chitin, is often present in fungal hyphae as the main component of the cell wall [40]. The toxicity of chitinases against fungi has been widely described [41]. However, data related to the direct action of these proteins against bacteria are rare. However, certain plant chitinases with additional lysozyme or lysozyme-like activity are able to cleave bacterial peptidoglycan [42]. In the present work, it was observed that conventional breeding of soybean cultivars produced a coordinated increase in several proteins involved in plant defense, which caused higher resistance in the Seridó-RCH cultivar. Efforts to engineer disease resistance in plants through the overexpression of PR proteins have shown that these proteins are not as effective when induced individually compared with when they are coordinately expressed [43].

Little attention has been paid to the possible food safety of new plant varieties derived from conventional breeding. However, the potential risks of the introduction or increased production of antinutritional or toxic compounds are applicable to the development of crop cultivars using both conventional breeding methods and transgenic plant technology [44]. To characterize the likelihood of unintended effects of the resistant cultivar (Seridó-RCH), with subsequent deleterious effects on humans and/or animals, a feeding trial was conducted in rats, using both cultivars as the sole source of protein. Diets were prepared using seeds produced under strict quality standards that were apparently free of pathologies related to microbial contamination and aflatoxins. Moreover, soybean seeds were not previously processed by autoclaving and/or irradiation, for example, to avoid changes, particularly in native protein constituents such as lectins, protease inhibitors and other proteins. Several studies have shown that autoclaving and irradiation alter the nutritional quality of certain proteins. Gamma ray exposition decreased the protein content of certain edible seeds in a dose-dependent manner, increasing the amount of free amino acids. In addition, this treatment decreased protein solubility, as observed for the albumin and globulin fractions of certain oil seeds (soybean, peanut and sesame) [45,46]. Furthermore, it is possible that radiation causes protein fragmentation and aggregation, with the latter effect leading to protein insolubility [47,48]. A protein-free (NPC) diet and an EW-based diet were used as controls. Although the daily food intake and amino acid composition were similar between the two cultivars, the rats fed with Seridó-RCH diets generally showed worse performance when protein quality indicators were compared.

A regression analysis relating the antinutritional and/or defense proteins to the nutritional parameters showed that lectin, trypsin inhibitor, peroxidase and chitinase were significant explanatory variables for the protein quality indicators. Lectin, trypsin inhibitor, peroxidase and chitinase significantly influence food intake in the reverse direction. However, weight gain is inversely influenced by lectin, trypsin inhibitor and chitinase, whereas peroxidase has no significant influence. NPU is also inversely influenced by lectin, trypsin inhibitor and chitinase. However, digestibility is significantly and inversely influenced by lectin and trypsin inhibitor but directly affected by chitinase. The antinutritional action of trypsin inhibitors and lectins is well known. Trypsin inhibitors decrease the availability of protein needed for optimal growth and health because they reduce the intestinal ability to digest protein, whereas lectins are known for their specific binding to cellular and intracellular membrane-associated carbohydrates (glycoprotein and glycolipids). Lectins also bind to carbohydrates in ingested food and limit or change their potential hydrolysis and absorption from the gut [49,50]. Regarding chitinase, the results showed that it contributed negatively to body weight gain as well as to NPU but not to digestibility. This phenomenon might have contributed to a lower biological value for the Seridó-RCH cultivar because this parameter may be calculated by dividing NPU by digestibility. These results, showing that chitinase is an antinutrient, are relatively surprising because nothing has ever been reported in this sense. Thus, further studies must be conducted to clarify this concept.

Although it was not possible to apply multiple regression analysis to the effects of the different proteins on organ weights, the diet based on Seridó-RCH, which exhibited higher activities for all of the studied proteins compared with Seridó, caused more prominent changes in certain organs, such as thymus atrophy and small intestine and pancreas enlargement. In previous work with other soybean cultivars, regression analysis data showed that small intestine enlargement was mainly induced by lectin, although trypsin inhibitor also had a significant effect. Conversely, trypsin inhibitor was the main factor responsible for pancreas enlargement, although lectin was also a significant explanatory variable. In the same study, urease was also a significant explanatory variable for pancreas and small intestine enlargement [4]. Thus, in the present study, it is probable that lectin, trypsin inhibitor and urease also contributed to alterations in these key organs.

5. CONCLUSIONS

Although they have the same protein contents, the non-isogenic soybean cultivars Seridó and Seridó-RCH, which are susceptible and resistant to stem canker, respectively, differed significantly in their levels of antinutritional and plant defense proteins. The increase in these proteins verified for Seridó-RCH was experimentally shown to cause a reduction in the protein quality of this resistant soybean cultivar. Although we cannot generalize our conclusion, the purpose of this work was to raise concerns about the possibility that conventional breeding processes for developing disease-resistant cultivars may bring about undesirable traits of this nature.

ACKNOWLEDGMENTS

This work was supported by the Brazilian Federal Agency for the Support and Evaluation of Graduate Education (CAPES, Brasilia, Brazil), the Brazilian National Council for Scientific and Technological Development (CNPq, Brasilia, Brazil) and Ceara State Research Foundation (FUNCAP, Fortaleza, Brazil).

REFERENCES

1. Ramanatha Rao, V.; Hodgkin, T. Genetic diversity and conservation and utilization of plant genetic resources. Plant Cell Tissue Organ Cult. **2002**, 68, 1–19.
2. Rischer, H.; Oksman-Caldentey, K.M. Unintended effects in genetically modified crops: Revealed by metabolomics? Trends Biotechnol. **2006**, 24, 102–104. [Google Scholar] [CrossRef] [PubMed]
3. Raboy, V. The future of crop breeding for nutritional quality. SABRAO J. Breed. Genet. **2013**, 45, 100–111.
4. Vasconcelos, I.M.; Maia, A.A.B.; Siebra, E.A.; Oliveira, J.T.A.; Carvalho, A.F.F.U.; Melo, V.M.; Carlini, C.R.; Castelar, L.I.M. Nutritional study of two Brazilian soybean (Glycine max) cultivars differing in the contents of antinutritional and toxic proteins. J. Nutr. Biochem. **2001**, 12, 55–62.
5. Gilani, G.S.; Cockell, K.A.; Sepehr, E. Effects of antinutritional factors on protein digestibility and amino acid availability in foods. J. AOAC Int. **1989**, 88, 967–987.
6. Grant, G. Antinutritional effects of soybean: A review. Prog. Food Nutr. Sci. (Oxford) **1989**, 13, 317–348.
7. Brune, M.F.S.S.; Pinto, M.O.; Peluzio, M.C.G.; Moreira, M.A. Barros, E.G. Biochemical and nutritional evaluation of a soybean line lacking the Kunitz trypsin inhibitor and lectins. Food Sci. Technol. **2010**, 30, 657–663.
8. Friedman, M.; Brandon, D.L.; Bates, A.H.; Hymowitz, T. Comparison of a commercial soybean cultivar and an isoline lacking the Kunitz trypsin inhibitor: Composition, nutritional value, and effects of heating. J. Agric. Food Chem. **1991**, 39, 327–335.
9. Liener, I.E. Implications of antinutritional components in soybean foods. Crit. Rev. Food Sci. Nutr. **1994**, 34, 31–67.
10. Haq, S.K.; Atif, S.M.; Khan, R.H. Protein proteinase inhibitor genes in combat against insects, pests, and pathogens: Natural and engineered phytoprotection. Arch. Biochem. Biophys. **2004**, 431, 145–159.
11. Vandenborre, G.; Smagghe, G.; Van Damme, E.J.M. Plant lectins as defense proteins against phytophagous insects. Phytochemistry **2011**, 72, 1538–1550.
12. Campelo, G.J.A.; Kiihl, R.A.S.; Almeida, L.A. Características Agronômicas e Morfológicas das Cultivares de Soja Desenvolvidas Para as Regiões de

Baixas Latitudes. Available online: http://www.cpatsa.embrapa.br/catalogo/livrorg/sojacultivares.pdf (accessed on 4 August 2013).
13. Tome, D. Criteria and markers for protein quality assessment—A review. Br. J. Nutr. **2012**, 108, S222–S229.
14. Vasconcelos, I.M.; Siebra, E.A.; Maia, A.A.B.; Moreira, R.A.; Neto, A.F.; Campelo, G.J.A.; Oliveira, J.T.A. Composition, toxic and antinutritional factors of newly developed cultivars of Brazilian soybean (Glycine max). J. Sci. Food Agric. **1997**, 75, 419–426.
15. Kakade, M.L.; Simons, N.R.; Liener, I.E. An evaluation of natural vs. synthetic substrates for measuring the antitryptic activity of soybean samples. Cereal Chem. **1969**, 46, 518–526.
16. Kaplan, A. The determination of urea, ammonia, and urease. Methods Biochem. Anal. **1969**, 17, 311–324.
17. Urbanek, H.; Kuzniak-Gebarowska, E.; Herka, K. Elicitation of defense responses in bean leaves by Botrytis cinerea polygalacturonase. Acta Physiol. Plant. **1991**, 13, 43–50.
18. Boller, T. Biochemical analysis of chitinases and β-1,3-glucanase. In Molecular Plant Pathology; Gurr, S.J., Mc Pherson, M.J., Bowles, D.J., Eds.; IRL Press: New York, NY, USA, 1993; pp. 23–29.
19. Pintér-Szakács, M.; Molnár-Perl, H. Determination of tryptophan in unhydrolyzed food and feedstuffs by the acid ninhydrin method. J. Agric. Food Chem. **1990**, 38, 720–726.
20. Milhome, M.A.L.; Lima, C.G.; de Lima, L.K.; Lima, F.A.F.; Sousa, D.O.B.; Nascimento, R.F. Occurrence of aflatoxins in cashew nuts produced in northeastern Brazil. Food Control **2014**, 42, 34–37.
21. Coates, M.E.; Odonogue, P.N.; Payne, P.R. Nutritional and microbiological recommendation. In Laboratory Animal Handbook 2—Dietary Standards for Laboratory Rats and Mice; Coates, M.E., Ed.; Laboratory Animals Limited: London, UK, 1969; pp. 13–15.
22. Miller, D.S.; Bender, A.E. The determination of the net utilization of protein by a shortened method. Br. J. Nutr. **1955**, 9, 382–389.
23. Triebold, H.O. Quantitative Analysis with Applications to Agricultural and Food Products; Van Nostrand Co.: New York, NY, USA, 1946.
24. Baethgen, W.E.; Alley, M.M. A manual colorimetric procedure for measuring ammonium nitrogen in soil and plant Kjeldahl digests. Commun. Soil Sci. Plant Anal. **1989**, 20, 961–969.
25. Morrison, D.F. Multivariate Statistical Methods; McGraw Hill Book Press: London, UK, 1978.

26. Mundlak, Y. On the concept of non-significant functions and its implications for regression analysis. J. Econom. **1981**, 16, 139–149.
27. WHO. Protein and Amino Acid Requirements in Human Nutrition—Report of Joint WHO/FAO/UNU Expert Consultation; World Health Organization Technical Report Series; WHO Press: Geneva, Switzerland, 2007; pp. 1–265.
28. Iqbal, A.; Khalil, I.A.; Ateeq, N.; Khan, M.S. Nutritional quality of important food legumes. Food Chem. **2006**, 97, 331–335.
29. Habib, H.; Fazili, K.M. Plant protease inhibitors: A defense strategy in plants. Biotechnol. Mol. Biol. Rev. **2007**, 2, 68–85.
30. Carrillo, L.; Herrero, I.; Cambra, I.; Sánchez-Monge, R.; Diaz, I.; Martinez, M. Differential in vitro and in vivo effect of barley cysteine and serine protease inhibitors on phytopathogenic microorganisms. Plant Physiol. Biochem. **2011**, 49, 1191–1200.
31. Kim, J.Y.; Park, S.C.; Hwang, I.; Cheong, H.; Nah, J.W.; Hahm, K.S.; Park, Y. Protease inhibitors from plants with antimicrobial activity. Int. J. Mol. Sci. **2009**, 10, 2860–2872.
32. De Hoff, P.L.; Brill, L.M.; Hirsch, A.M. Plant lectins: The ties that bind in root symbiosis and plant defense. Mol. Genet. Genomics **2009**, 282, 1–15.
33. Charungchitrak, S.; Petsom, A.; Sangvanich, P.; Karnchanatat, A. Antifungal and antibacterial activities of lectin from the seeds of Archidendron jiringa Nielsen. Food Chem. **2011**, 126, 1025–1032.
34. Real-Guerra, R.; Stanisçuaski, F.; Carlini, C.R. Soybean urease: Over a hundred years of knowledge. In Comprehensive Survey of International Soybean Research—Genetics, Physiology, Agronomy and Nitrogen Relationships, 1st ed.; Board, J.E., Ed.; InTech: Dubrovnik, Croatia, 2013; pp. 317–339.
35. Becker-Ritt, A.B.; Martinelli, A.H.; Mitidieri, S.; Feder, V.; Wassermann, G.E.; Santi, L.; Vainstein, M.H.; Oliveira, J.T.A.; Fiuza, L.M.; Pasquali, G.; et al. Antifungal activity of plant and bacterial ureases. Toxicon **2007**, 50, 971–983.
36. Wiebke-Strohm, B.; Pasquali, G.; Margis-Pinheiro, M.; Bencke, M.; Bücker-Neto, L.; Becker-Ritt, A.B.; Martinelli, A.H.S.; Rechenmacher, C.; Polacco, J.C.; Stolf, R.; et al. Ubiquitous urease affects soybean susceptibility to fungi. Plant Mol. Biol. **2012**, 79, 75–87.

37. Stam, J.M.; Kroes, A.; Li, Y.; Goes, R.; van Loon, J.J.A.; Poelman, E.H.; Dicke, M. Plant interactions with multiple insect herbivores: From community to genes. Annu. Rev. Plant Biol. **2014**, 65, 689–713.
38. Cipollini, D.; Wang, Q.; Whitehill, J.G.A.; Powell, J.R.; Bonello, P.; Herms, D.A. Distinguishing defensive characteristics in the phloem of ash species resistant and susceptible to emerald ash borer. J. Chem. Ecol. **2011**, 37, 450–459.
39. Wang, S.; Shao, B.; Rao, P.; Deng, Z.; Xie, M. Limlin, a novel leguminous peroxidase with antifungal activity from Phaseolus limensis. J. Food Biochem. **2011**, 35, 1206–1222.
40. Vega, K.; Kalkum, M. Chitin, chitinase responses, and invasive fungal infections. Int. J. Microbiol. **2012**.
41. Wang, S.; Ye, X.; Chen, J.; Rao, P. A novel chitinase isolated from Vicia faba and its antifungal activity. Food Res. Int. **2012**, 45, 116–122.
42. Schultze, M.; Staehelin, C.; Brunner, F.; Genetet, I.; Legrand, M.; Fritig, B.; Kondorosi, E.; Kondorosi, A. Plant chitinase/lysozyme isoforms show distinct substrate specificity and cleavage site preference towards lipochitooligosaccharide Nod signals. Plant J. **1998**, 16, 571–580.
43. Collinge, D.B.; Jorgensen, H.J.L.; Lund, O.S.; Lyngkjaer, M.F. Engineering pathogen resistance in crop plants: Current trends and future prospects. Annu. Rev. Phytopathol. **2010**, 48, 269–291.
44. Paparini, A.; Romano-Spica, V. Public health issues related with the consumption of food obtained from genetically modified organisms. Biotechnol. Annu. Rev. **2004**, 10, 85–122.
45. Maity, J.P.; Chakraborty, S.; Kar, S.; Panja, S.; Jean, J.S.; Samal, A.C.; Chakraborty, A.; Santra, S.C. Effects of gamma irradiation on edible seed protein, amino acids and genomic DNA during sterilization. Food Chem. **2009**, 114, 1237–1244.
46. El-Niely, H.F.G. Effect of radiation processing on antinutrients, in vitro protein digestibility and protein efficiency ratio bioassay of legume seeds. Radiat. Phys. Chem. **2007**, 76, 1050–1057.
47. Kumta, U.S.; Tappel, A.L. Radiation damage to proteins. Nature **1961**, 191, 1304–1305.
48. Afify, A.E.M.R.; Rashed, M.M.; Mahmoud, E.A.; El-Beltagi, H.S. Effect of gamma radiation on protein profile, protein fraction and solubility's of three oil seeds: Soybean, peanut and sesame. Not. Bot. Horti Agrobot. Cluj-Napoca **2011**, 39, 90–98.

49. Muzquiz, M.; Varela, A.; Burbano, C.; Cuadrado, C.; Guillamón, E. Bioactive compounds in legumes: Pronutritive and antinutritive actions. Implications for nutrition and health. Phytochem. Rev. **2012**, 11, 227–244.
50. Kumar, S.; Verma, A.K.; Das, M.; Jain, S.K.; Dwivedi, P.D. Clinical complications of kidney bean (Phaseolus vulgaris L.) consumption. Nutrition **2013**, 29, 821–827.

6
CHAPTER

COMPARATIVE PROTEOMIC ANALYSIS OF COTTON FIBER DEVELOPMENT AND PROTEIN EXTRACTION METHOD COMPARISON IN LATE STAGE FIBERS

Hana Mujahid [1], Ken Pendarvis [2,†], Joseph S. Reddy [3,‡], Babi Ramesh Reddy Nallamilli [1,§], K. R. Reddy [4], Bindu Nanduri [3] and Zhaohua Peng [1,*]

[1]*Department of Biochemistry, Molecular Biology, Entomology, and Plant Pathology, Mississippi State University, Mississippi State, MS 39762, USA*

[2]*Institute for Genomics, Biocomputing and Biotechnology, Mississippi Agricultural and Forestry Experiment Station, Mississippi State University, Mississippi State, MS 39762, USA*

[3]*College of Veterinary Medicine, Mississippi State University, Mississippi State, MS 39762, USA*

[4]*Department of Plant and Soil Sciences, Mississippi State University, Mississippi State, MS 39762, USA*

ABSTRACT

The distinct stages of cotton fiber development and maturation serve as a single-celled model for studying the molecular mechanisms of plant cell elongation, cell wall development and cellulose biosynthesis. However, this model system of plant cell development is compromised for proteomic studies due to a lack of an efficient protein extraction method during the later stages of fiber development, because of a recalcitrant cell wall and the presence of abundant phenolic compounds. Here, we compared the quality and quantities of proteins extracted from 25 dpa (days post anthesis) fiber with multiple protein extraction methods and present a comprehensive quantitative proteomic study of fiber development from 10 dpa to 25 dpa. Comparative analysis using a label-free quantification method revealed 287 differentially-expressed proteins in the 10 dpa to 25 dpa fiber developmental period. Proteins involved in cell wall metabolism and regulation, cytoskeleton development and carbohydrate metabolism among other functional categories in four fiber developmental stages were identified. Our studies provide protocols for protein extraction from maturing fiber tissues for mass spectrometry analysis and expand knowledge of the proteomic profile of cotton fiber development.

Keywords: cotton; fiber elongation; cell wall; comparative proteomics; label-free

1. INTRODUCTION

Being the world's leading natural textile fiber, the economically-valuable cotton fibers of upland cotton (Gossypium hirsutum L.) are unique in the plant kingdom due to their size and chemical composition [1,2]. Each cotton fiber is a single and long (≥2.25 cm) cell originating from the ovule epidermis [2,3,4,5]. Developing near-synchronously during seed development, cotton fiber development consists of four overlapping developmental stages: fiber initiation, cell elongation (primary cell wall synthesis), cell wall thickening (secondary cell wall deposition) and maturation [2,3,5,6]. During primary wall elongation (within 20 days post anthesis (dpa)) and secondary wall deposition and thickening (from 20 to 35 dpa), sizeable amounts of polysaccharide components are synthesized and deposited, producing a cell wall 3 to 4 μm thick, made up of more than 94% cellulose [3,7,8]. These advantageous features make cotton fiber an excellent single-celled model for studying the molecular mechanisms of plant cell elongation, cell wall development and cellulose biosynthesis [2,5,9,10,11].

Progress has been made in the large-scale identification of genes and proteins involved in cotton fiber elongation in the last decade [9,12,13]. Several comparative proteome and transcriptome studies during different stages of cotton fiber development have been reported [2,5,8,11,12]. In addition, the transcriptome and proteome comparisons between reduced fiber and fiberless mutants and their respective parental wild-types (WT) have also been conducted [6,14,15,16,17,18,19,20,21,22]. However, due to the recalcitrant nature of cotton fiber, most of the reported studies used young fiber tissue as experimental materials instead of fibers in later developmental stages. Initially, cotton fiber proteins were extracted by directly homogenizing cotton fibers in aqueous buffer followed by organic solvent precipitation [7,23]. However, this method was unsuitable for two-dimensional gel electrophoresis due to the horizontal and vertical streaking and smearing caused by the phenolic and other contaminants co-extracted with proteins [7,23]. In recent years, cotton fiber proteins have been mainly extracted with modifications on the phenol-based procedure [2,5,7,8,10,11,13,20,22,24,25] and by the trichloroacetic acid extraction method, as described by Pang et al. [6]. However, there have not been any reports of the successful extraction of cotton fiber proteins for shotgun proteomics from maturing fiber tissues, for example fiber stages after 30 dpa. Despite all of the reported studies, the underlying mechanisms behind fiber initiation, elongation and maturation are still largely unknown [13,21].

The development of novel strategies that optimize protein extraction for cotton fiber cells, particularly the stages after 25 dpa, is critical for using mass spectrometry-based proteomic approaches to study cotton fiber development. Pressure cycling technology (PCT) uses a specifically designed device (Barocycler™) and reaction containers (PULSE™ tubes) to apply cycles of hydrostatic pressure to samples [26,27]. PCT provides a simple, fast, effective and reproducible process to release cellular contents from biological samples [26,27,28]. Previously, it has been shown that the use of PCT increased protein yields from E. coli, where PCT extracted 14.2% more total protein than using a standard bead mill [29,30]. Furthermore, 2-DE showed 801 protein spots in the PCT lysate, compared to 760 spots in the bead mill lysate [29]. In mammalian liver tissue, PCT isolated more protein, as well as unique proteins when compared to protein isolation using a Polytron or ground-glass (GG) homogenizers [30,31]. Szabo et al. found that PCT-assisted glycan release resulted in the rapid release of asparagine-linked glycans from bovine ribonuclease B, human transferrin and polyclonal human immunoglobulin [32]. It is thought that high pressure alters the protein conformation, pushing water molecules into the protein interior, thus leading to protein unfolding [32,33]. Furthermore, for heat-sensitive molecules, PCT provides an advantage by being able to be conducted at mild temperatures (room temperature to 37 °C) [32,34]. Szabo et al. showed that PCT offers several advantages, including not causing decomposition (e.g., desialylation) of the glycan structures, the speed of extraction and the ability to simultaneously process 12 samples at a time [32]. However, the PCT effect on plant protein extraction, particularly the recalcitrant samples, such as cotton fiber, remains to be examined.

In this report, we identified 1446 proteins in four time points of fiber development (10 dpa, 15 dpa, 25 dpa and 35 dpa). Comparison of the proteomes of different stages of fiber development revealed 287 differentially-regulated proteins, functionally involved in cytoskeleton development, energy/carbohydrate metabolism and cell wall development, among other processes. This study presents the first proteome of the 35 dpa cotton fiber, a highly recalcitrant tissue, and presents a comprehensive proteomic study of fiber development from 10 dpa to 25 dpa. In addition, it provides protocols for protein extraction from maturing fiber tissues for mass spectrometry analysis, and our results considerably expand the knowledge of the cotton fiber proteome during development.

2. EXPERIMENTAL SECTION

2.1. Plant Materials

Upland cotton (Gossypium hirsutum L.) cultivar Texas Marker (TM)-1 was grown in sunlit, controlled environment chambers, known as soil-plant-atmosphere-research (SPAR) units, located at the Rodney Foil Plant Science Research Center, Mississippi State University, Mississippi, USA, consisting of fine sand as the growing medium [35].

2.2. Growth Conditions

Detailed operations and controls of SPAR chambers have been previously described [35]. Briefly, the SPAR chamber consisted of a steel soil bin (1 m deep × 2 m long × 0.5 m wide) to accommodate the root system, a Plexiglas chamber (2.5 m tall × 2 m long × 1.5 m wide) to accommodate aerial plant parts and a heating and cooling system connected to air ducts that passed conditioned air to cause leaf flutter through the plant canopy. Variable density shade cloths placed around the edges of the plant canopy, designed to simulate canopy spectral properties, were adjusted regularly to match canopy height and to eliminate the need for border plants. Four rows with five plants per row were maintained in the chamber until harvest. A day/night temperature of 30/22 °C was maintained throughout the experiment. The temperature control was achieved to the desired set points using chilled ethylene glycol supplied to the cooling system via several parallel solenoid valves that were opened and closed depending on the cooling requirements, an electrical resistance heater, which provided short pulses of heat, and a fan, which provided air circulation throughout the chamber [35]. A carbon dioxide concentration of 400 ppm was maintained during the experiment. CO_2 concentration was monitored and adjusted every 10 s throughout the day and maintained at 400 ± 10 μL·L^{-1} during the daylight hours using a dedicated LI-6250 CO_2 analyzer (Li-COR, Inc., Lincoln, NE, USA). Plants were well watered with full-strength Hoagland's nutrient solution [36] three times a day through a programmed drip irrigation system in order to maintain optimum water and nutrient supply throughout the experiment. The SPAR units were supported by an environmental monitoring and control system, which provided automatic acquisition and storage of data, monitored every 10 s throughout the day and night. During the experiment, the incoming daily solar radiation (285 to 2800 nm) outside of the SPAR units was measured with a pyranometer (Model 4–8; The Eppley Laboratory Inc., Newport, RI, USA), which ranged from 1.4 to 27.2 MJ·m^{-2}·d^{-1} with an average of 15.6 MJ·m^{-2}·d^{-1}. The relative humidity (RH) of the chamber was monitored with a humidity sensor (HMV 70Y, Vaisala Inc., San Jose, CA, USA) installed in the returning path of airline ducts. The average relative humidity during the experimental period was 34.5% ± 3.0%.

Cotton bolls were labeled and tagged on the anthesis day. Harvested fiber time points for protein extraction were 10 days post anthesis (dpa), 15 dpa, 25 dpa and 35 dpa. Fiber samples were collected by hand, immediately frozen in liquid nitrogen and stored at −80 °C until protein extraction was performed.

2.3. Protein Extraction with Phenol

Protein extraction with phenol was performed as previously described [24,37,38,39]. Briefly, 2 grams of cotton fiber were ground under liquid nitrogen using a mortar-pestle; the tissue was suspended in phenol extraction buffer (PEB) containing 0.9 M sucrose, 0.5 M Tris-Cl, 0.05 M EDTA, 0.1 M KCl and 2% 2-mercaptoethanol (added freshly), pH 8.7, and homogenized well. The same volume of phenol was added to the sample, and the sample was homogenized well. The phenol phase was collected by centrifugation at 7000 rpm for 15 min at 4 °C, and phenol extraction was repeated three times. Five volumes of precipitation solution (methanol, 0.1 M ammonium acetate and 1% 2-mercaptoethanol (added freshly)) were added to the final collected phenol phase. To precipitate, the sample was stored at −70 °C overnight. To collect the precipitate, the sample was centrifuged at 12,000 rpm for 15 min at 4 °C. The pellet was collected and washed three times with 70% ethanol. The pellet was vacuum dried and stored at −80 °C. Three replicates were included for extraction methods comparison for the 25 dpa samples and three replicates for comparison were included for differential expression analyses of all of the samples involved.

2.4. Protein Extraction with Phenol and PCT for 25 dpa Fiber

Approximately 400 mg of cotton fiber were suspended in PEB buffer with phenol. Tissue was disrupted using the PBI PCT shredder™ following the manufacturer's instructions, followed by PCT using the Barocycler® NEP2017 pressure cycling instrument (Pressure BioSciences Inc., South Easton, MA, USA) in PCT Shredder PULSE Tubes™ for 60 cycles at 35,000 psi [29,40]. After PCT completion, the remaining phenol extraction steps as described in Section 2.3 were performed.

2.5. Protein Extraction with IEF Reagent for 25 dpa Fiber

Approximately 400 mg of cotton fiber were ground under liquid nitrogen. After tissue disruption, fiber was suspended in isoelectric focusing (IEF) buffer (7 M urea, 2 M thiourea, 4% 3-[(3-cholamidopropyl)dimethylammonio]-1-propanesulfonate (CHAPS)) and vortexed well. After vortexing, samples were centrifuged, and the supernatant was collected.

2.6. Protein Extraction with IEF Reagent and PCT for 25 dpa and 35 dpa Fiber

Approximately 400 mg of cotton fiber were suspended in IEF buffer, followed by PCT, as described in Section 2.4.

2.7. Protein Digestion

Protein concentration was determined using the Bradford assay, and equal amounts of proteins were prepared for digestion. Protein digestion was carried out as previously described [38,39,41,42]. After desalting the digested peptides, the eluted peptides were dried by vacuum and dissolved in 5% acetonitrile (ACN) and 0.1% formic acid. The injection volume for mass analysis was twenty microliters (50 μg digested peptides).

2.8. Shotgun Proteomic Analysis

Chromatography equipment consisted of a Thermo Surveyor HPLC system operated at 500 nL per minute via a split solvent line. A gradient from 5% to 95% acetonitrile in 600 min was used for peptide separation, followed by a 25-min hold at 95% acetonitrile and a 30 min column re-equilibration period; all solvents contained 0.01% formic acid as an ion source. Ten days post anthesis (dpa), 15 dpa and some 25 dpa samples were separated using this gradient for differential expression analysis, while the remaining 25 dpa samples (for method comparison) and 35 dpa samples were analyzed using a 5% to 50% gradient with identical wash and equilibration steps. The latter gradient provided better peptide resolution and was employed in a previous publication [39]. A 0.75 mm by 100 mm BioBasic C18 column

(Thermo 72105–100266) was used for peptide separation using both gradients. The Surveyor was coupled with a Thermo LCQ DECA XP Plus mass spectrometer with a stock nanospray ion source. Data were collected over a total duration of 655 min for each sample using MS scans directly followed by three tandem MS/MS scans on the three most intense precursor masses from the full MS scan. Dynamic mass exclusion windows were 2 min long with a repeat count of two. All data collection parameters were identical for both HPLC gradients.

2.9. Protein Identification and Statistical Analysis

Spectral files (RAW) were converted to mgf format using msConvert (available in the ProteoWizard tool kit) [43] and searched against the Gossypium database downloaded on 22 April 2015 from UniProt, using X!tandem [44]. The database contained 37,423 entries. X!tandem was configured to allow a maximum of 2 missed tryptic cleavage sites and the precursor and fragment mass tolerance were set to 1000 and 500 ppm, respectively. Amino acid modifications included in the database search were single and double oxidation of methionine, carboxymethylation and carbamidomethylation of cysteine, methylation of arginine and lysine residues, phosphorylation of serine and threonine residues, as well as spontaneous water loss after phosphorylation of serine and threonine residues. X!tandem results, by default, consist of individual peptide-spectrum matches, with each match containing information about the corresponding peptide sequence and parent protein. In order to view the results on the protein level, with subsequent peptide sequences listed as a group, they needed to be reorganized. This was accomplished using the Perl programming language to parse the X!tandem results files. As spectrum matches were processed, those with an E-value greater than 0.05 were discarded and the remainder organized by corresponding protein. Proteins identified by a single peptide match were also discarded. In order to evaluate the quality of the dataset (false discovery rate), decoy searches were performed using a randomized version of the Gossypium protein database and processed using the same Perl logic as before. For all protein search results files, the false discovery rate was less than 0.01 (less than 1%). Results passing all criteria were subsequently organized by dpa and extraction method using Perl. Protein coverage was calculated using Perl. For each protein sequence, an array of zeros the same length as the protein was created. Peptides were string-matched to protein sequence and the coordinates recorded. Elements of the array having the same coordinates were set equal to 1. Once all peptides were processed, the elements were summed, divided by the protein length and multiplied by 100 to calculate coverage.

2.10. Protein Grouping

Since the database used for protein identification contained several species of Gossypium, the filtered results included several orthologs identified with the same set of peptides. To reduce redundancy in protein identification, protein grouping was performed using in-house Perl scripts [45] with the following conditions: if two or more proteins were identified with an identical set of spectra in Gossypium, only one protein ID (preferentially from Gossypium hirsutum) was retained in the protein list, and the remaining protein IDs were grouped together; if two or more proteins were identified with an identical set of spectra in Gossypium hirsutum, they were retained in order to account for isoforms in Gossypium hirsutum. Only proteins retained after grouping were considered for further analysis. Each group and associated proteins can be found in Supplemental Table S2.

2.11. Protein Quantification and Statistical Analysis

To evaluate differences between dpa samples on a proteome level, a differential expression analysis based on peptide elution profiles was performed using Perl. Precursor mass spectra were extracted from the raw data in MS1 format using the msConvert tool from the ProteoWizardtoolkit [43]. Peptide precursor m/z values were extracted from the previously compiled protein identifications. Elution profiles for peptide-spectrum matches were calculated by parsing each corresponding MS1 file and summing the ion current for that match's m/z value within a 1-Da tolerance, effectively integrating the elution profiles. Each trace started at the scan number of the peptide-spectrum match and proceeded both forward and backward until the chromatogram noise level, or a distance of 250 scans, was reached. Multiple peptide-spectrum matches with the same precursor m/z were only counted once, ensuring the same integral was not counted multiple times. Once all peptide-spectrum matches were processed, intensities were summed for each protein on a per-replicate basis. Proteins not identified in a replicate were represented with the average noise level of the replicate's chromatogram for further calculations. The reasoning behind this is two-fold: (1) peptides not identified in a replicate could be present at levels at or below the noise level of the chromatogram, causing the mass spectrometer to ignore them; and (2) for calculating expression ratios between lines, zero cannot be in the denominator. Data were normalized using a mode-based technique. First, the mode of the protein intensities for each replicate was calculated, representing the most commonly-occurring protein intensity. Next, for each identified protein, the intensity per replicate was divided by the mode of the same replicate. This ensures that normalization is not affected by the minimum and maximum intensities, which can vary tremendously between replicates. For each protein, the ratio between lines was calculated from the replicate intensities. A Monte Carlo resampling analysis was performed to evaluate the significance of the intensity distribution. For each replicate, a random intensity was generated between and including the minimum noise level and maximum intensity across all replicates. A new ratio between lines was calculated and compared to the original. This process was performed one million times, and the number of times the random ratio was above or below the experimental ratio was recorded. From this, a p-value was calculated for each distribution to indicate significance. Expression differences with a p-value ≤ 0.05 were considered significantly differentially expressed.

3. RESULTS

3.1. Comparison of Different Protein Extraction Buffers and Tissue Grinding Methods

Cotton protein extraction for proteomics studies has largely been performed using modifications on the traditional phenol extraction method as reported [2,5,7,8,10,11,13,20,22,24,25] and TCA extraction as described by Pang et al. [6]. As cotton fiber tissue matures, however, the protein content decreases, the cell wall thickens and interfering content, such as polysaccharides, polyphenolic compounds, pectin, lipids and waxes, is present, which substantially affect downstream proteomic applications [7,46]. To optimize the protein extraction method for cotton fiber, particularly the fiber from later developmental stages, such as 25 dpa and 35 dpa, we examined different extraction procedures using 25 dpa fibers as shown in Figure 1.

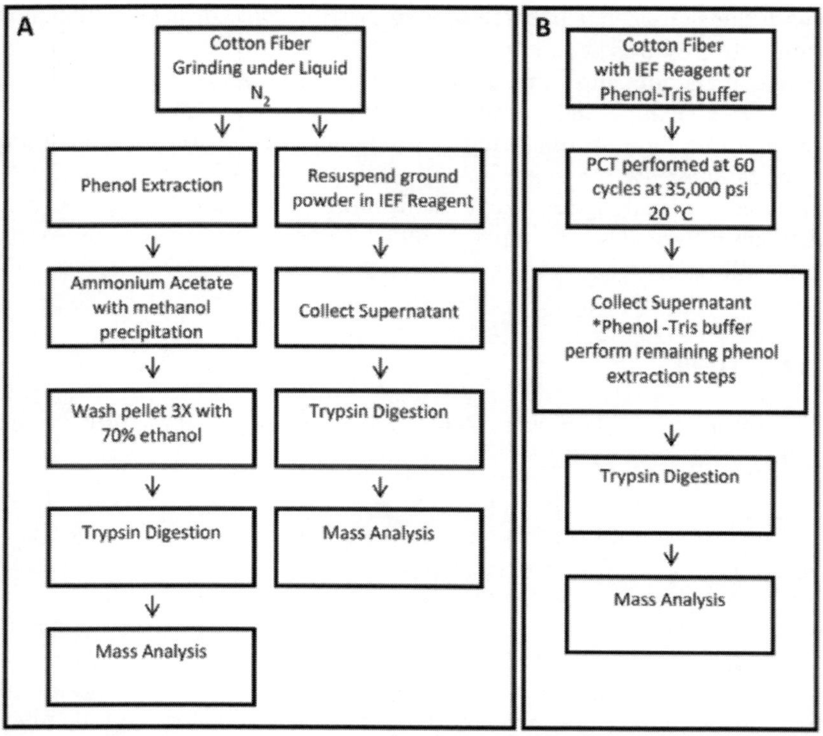

Figure 1. Summary of protein extraction methods used in this study. Cotton fiber proteins from 25 days post anthesis (dpa) fiber were extracted using different protein extraction methods, as shown in (**A**) and (**B**). Cotton fiber proteins were extracted with the mortar-pestle method combined with isoelectric focusing IEF reagent (A), the mortar-pestle method combined with phenol-Tris buffer (A) and with pressure cycling technology (PCT) combined with IEF reagent or phenol-Tris buffer (B). *After PCT combined with phenol-Tris buffer was completed, remaining phenol extraction steps were performed as described in Section 2.3.

We found that pairing the pressure cycling technology with both the IEF buffer and phenol could improve protein identification. Grinding the sample under liquid nitrogen (mortar-pestle extraction) and using the IEF buffer identified 845 proteins. When PCT was used with the IEF buffer, 884 proteins were identified, from which 235 were newly-identified proteins (Figure 2). When phenol extraction was coupled with PCT, 980 proteins were identified, while grinding the sample under liquid nitrogen and using phenol alone identified 934 proteins (Figure 2).

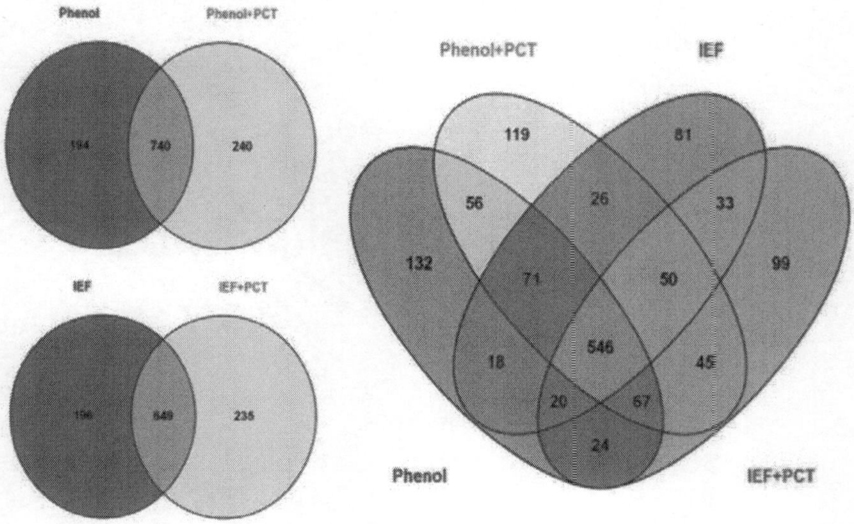

Figure 2. Analysis of 25 dpa cotton fiber proteome obtained using four different extraction methods (phenol, phenol + PCT, IEF and IEF + PCT). Venn diagrams displaying the number of identified cotton proteins in each extraction procedure for 25 dpa cotton fiber and the overlap of identified proteins among extraction procedures. Venn diagrams were generated using the Venny 2.0 tool [47].

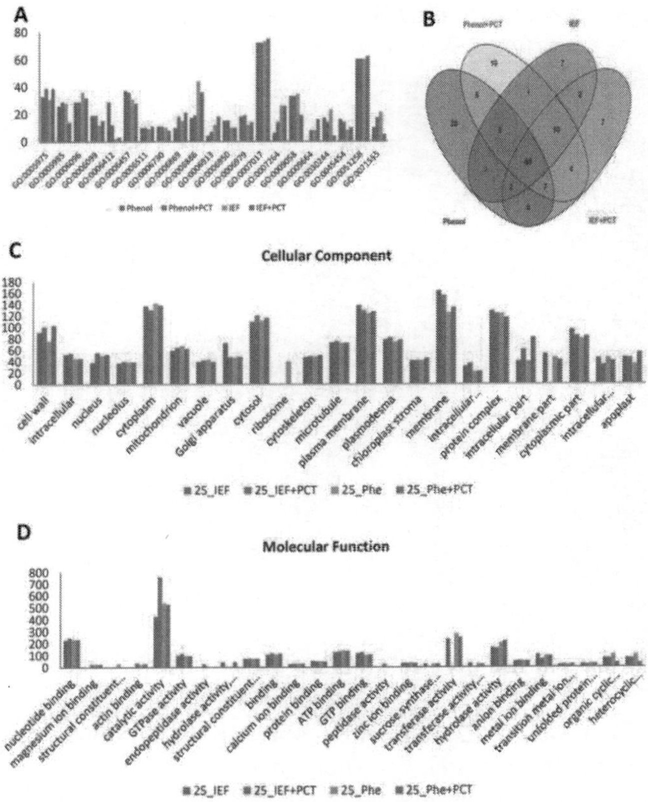

Figure 3. Gene ontology (GO) annotations of proteins identified using four protein extraction procedures in 25 dpa cotton fiber. Protein number, y-axis; gene ontology terms, x-axis. Phe, phenol. Phe+PCT, phenol+PCT. Only predominant GO category terms are shown in figure (**A**). Predominant biological process GO category terms of 25 dpa fiber proteins identified by the four extraction procedures, respectively. (**B**) Venn diagram displaying overlapping biological process category terms across four extraction procedures used for 25 dpa fiber. (**C**) Predominant cellular component category terms for proteins identified in 25 dpa fiber using four extraction methods, respectively. (...) in the x-axis indicates an abbreviated term; complete names of abbreviated cellular component GO terms from left to right include: intracellular membrane-bounded organelle || intracellular organelle part. (**D**) Predominant molecular function category terms for proteins identified in 25 dpa fiber using four extraction methods, respectively. (...) in the x-axis indicates abbreviated terms; complete names of abbreviated molecular function GO terms from left to right include: structural constituent of ribosome || hydrolase activity, hydrolyzing O-glycosyl compounds || structural constituent of cytoskeleton || sucrose synthase activity || transferase activity, transferring glycosyl groups || transition metal ion binding || unfolded protein binding || organic cyclic compound binding || heterocyclic compound binding.

For functional classification of the identified proteins, we carried out gene ontology analysis using the UniProt-GOA database (www.uniprot.org). The proteins were classified by gene ontology annotation based on three categories: biological process, cellular component and molecular function. Analysis of the identified proteins by each of the four extraction methods showed that the majority of the proteins identified in all methods were involved in microtubule-based processes (GO:0007017) and protein polymerization (GO:0051258) (Figure 3A). The majority of the GO terms from these categories overlapped significantly across the four methods, with phenol identifying the most unique biological processes' GO terms(Figure 3B). Using the GOSlimAuto [48] available at AgBase, the cellular component and molecular function gene ontology terms were compared across the four protein extraction procedures (Figure 3C,D). Although the majority of proteins identified across the four extraction methods did not significantly belong to any one category, a large number did come from the cytoplasm, cytosol, cell wall, plasma membrane, protein complex and membrane (Figure 3C). Interestingly, ribosomal proteins were only identified in the phenol extracted samples (Figure 3C). When looking at the molecular function, the majority of the proteins had catalytic activity and nucleotide binding (Figure 3D).

3.2. Fiber Proteome

We identified 451 proteins with two or more peptides in 10 dpa cotton fiber using the phenol extraction method. Meanwhile, 407 proteins were identified with two or more peptides in the 15 dpa fibers using the same extraction method. For 25 dpa fiber, a total of 1387 proteins were identified with two or more peptides when all four extracting methods were included (Figure 2). In an effort to improve protein identification for later stage fibers, both IEF + PCT and phenol extraction methods were used for 35 dpa fiber. Using these two methods, a total of 785 proteins were identified with two or more peptides. With all protein extraction methods combined in this study, we identified 1446 proteins of the fiber proteome (Supplemental Table S1).

3.3. Functional Analysis of the 35 dpa Proteome

Functional analysis of the total identified proteins in the 35 dpa fiber proteome revealed a strong role of cytoskeletal and energy/carbohydrate metabolism proteins at this fiber stage (Supplementary Figures S1–S3). Major biological processes involving 35 dpa proteins included the metabolic process, oxidation-reduction process, microtubule based process, seed trichome elongation, protein polymerization, response to stress, carbohydrate metabolic process, transport, etc. (Supplementary Figure S1).

The 35 dpa proteome contained proteins from the cytoplasm, plasma membrane, protein complex, membrane, cytosol, cell wall, plasmodesma, cytoplasmic part, microtubule, mitochondrion, Golgi apparatus, etc. (Supplementary Figure S2). The predominant molecular functions of proteins identified in the 35 dpa proteome included catalytic activity, nucleotide binding, transferase activity, hydrolase activity, ATP binding, GTP binding, GTPase activity, structural constituent of cytoskeleton, metal ion binding, etc. (Supplementary Figure S3). The proteins in the 35 dpa proteome belonged to a broad range of protein families (Figure 4). The major protein families identified in the 35 dpa proteome included the tubulin family, actin family, zinc containing alcohol dehydrogenase family, 14-3-3 family, heat shock protein 70 family, annexin family, plant LTP family, profilin family, ATPase alpha/beta chains family, etc. (Figure 4). Cytoskeletal proteins identified in the 35 dpa proteome included several actin isoforms, as well as alpha and beta-tubulin isoforms and annexin, among others. Energy/carbohydrate metabolism proteins included enolase isoforms, fructokinase (D2D2Z5), glyceraldehyde-3-phosphate dehydrogenase, sucrose synthase 1, 4, 2 and sucrose synthase isoforms B, C and D. The presence of proteins involved in cytoskeletal and carbohydrate metabolism in the 35 dpa proteome is understandable, since cellulose is the major carbohydrate in cotton fiber at maturity [49]. A large amount of energy is needed to produce and deposit the secondary wall cellulose microfibrils [49].

The 35 dpa proteome also contained many redox-related proteins. Proteins in this category included 2-nitropropane dioxygenase (D2D302), 3-ketoacyl-CoA thiolase 2 (A0A0B0N1S6), ascorbate peroxidase (C6ZDA9), protein disulfide isomerase (G1ED19), benzoquinone reductase (A3F7Q2), superoxide dismutase (Q3SAW9), catalase isozyme 1, 2 (A0A0B0PAB0, P30567), among others. Several cell wall synthesis-related proteins were also identified, including endo-1,3-beta-glucanase (O23953), several endo-alpha-1,4-glucanase isoforms (F1BWY1, Q4F885, F1BWY2), pectin methylesterase 5 (R9QQU5), UDP glucose pyrophosphorylase (Q6RY01), GRP-like protein 2 (Q0PW29) and fasciclin-like arabinogalactan protein 6, 2, 5, 1 and 4 (A9XTL1, A9XTK7, A9XTL0, B1NHU5, A9XTK9), among others.

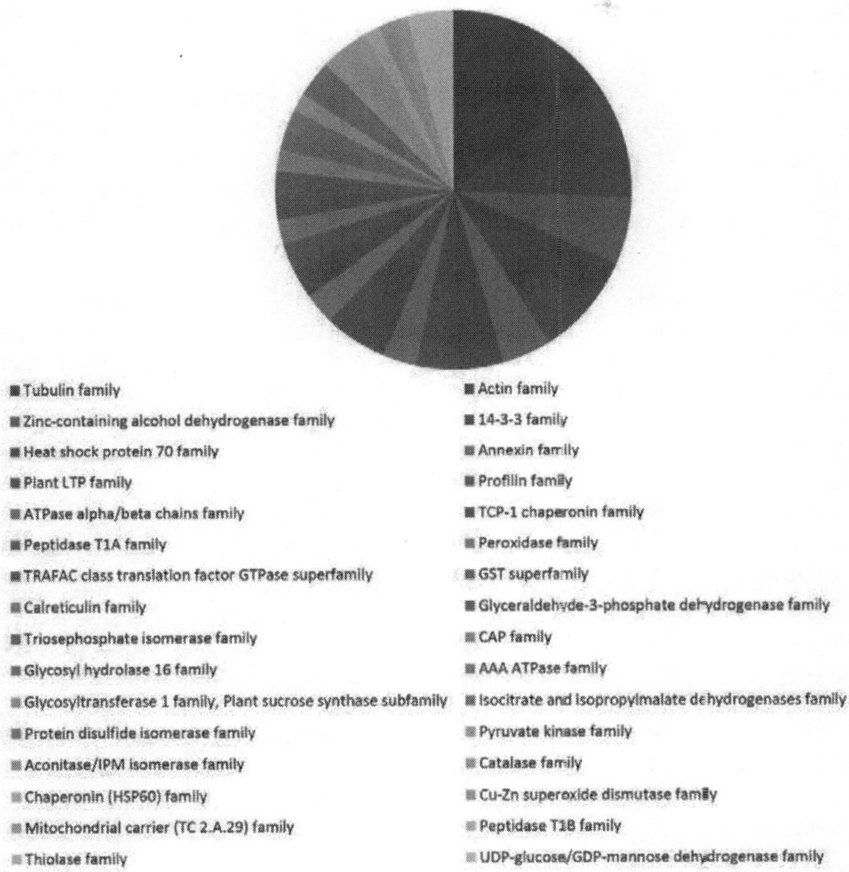

Figure 4. Predominant protein families identified in the 35 dpa cotton fiber proteome. The section size is proportional to the numbers of proteins identified in the category. The name of each category is shown below the chart.

3.4. Differentially-Expressed Proteins during Fiber Development

During fiber development, fiber cells undergo a series of morphological and compositional changes. To further examine developmental regulation of the cotton fiber proteome, we examined differentially-expressed proteins in different stages of fiber development. We quantitatively compared the proteome changes in the following stages: 15 dpa vs. 10 dpa and 25 dpa vs. 15 dpa. A non-labeling quantification method based on peptide spectral intensity was used [50]. Differential expression was only considered for proteins with a p-value ≤ 0.05.

Following the course of fiber development, we found that 287 proteins were differentially expressed in total (Supplemental Table S3). To further analyze the differentially-regulated proteins, functional classification of the differentially-expressed cotton proteins was carried out according to the gene ontology (GO) rules using GOSlimAuto available at AgBase [39,48,51,52].

Functional analysis and comparison of the biological processes of the upregulated proteins showed that when 15 dpa vs. 10 dpa stages were compared, the predominant biological processes were the oxidation-reduction process, biosynthetic process and sucrose metabolic process, among others (Supplementary Figure S4). The predominant categories of the cellular component of the upregulated proteins in this comparison (15 dpa vs. 10 dpa) were cytosol, intracellular and nucleus, among others (Supplementary Figure S5). The predominant molecular function categories of the comparison (15 dpa vs. 10 dpa) were oxidoreductase activity, metal ion binding and zinc ion binding, among others (Supplementary Figure S6). When 25 dpa vs. 15 dpa stages were compared, the predominant biological processes of the upregulated proteins were the metabolic process, microtubule-based process and protein polymerization, among others (Supplementary Figure S4); the predominant cellular components were microtubule, cytoplasm and cytosol, among others (Supplementary Figure S5); and the predominant molecular function categories were nucleotide binding, GTPase activity and structural constituent of cytoskeleton (Supplementary Figure S6) in the same comparison.

When 15 dpa vs. 10 dpa downregulated proteins were compared, proteins belonging to biological processes, such as metabolic process, transport and plant-type cell wall organization, among others, were the predominant (Supplementary Figure S7). Meanwhile, proteins in membrane, cytoplasm and cell wall were predominant categories of cellular components in the same comparison of the downregulated proteins. (Supplementary Figure S8). Additionally, proteins in catalytic activity, transferase activity and hydrolase activity, among others, were the main molecular function categories in the comparison (15 dpa vs. 10 dpa) (Supplementary Figure S9). The 25 dpa vs. 15 dpa downregulated protein comparison showed a predominance of proteins involved in the oxidation-reduction process, transport and protein folding, among others (Supplementary Figure S7). The 25 dpa vs. 15 dpa downregulated protein comparison showed a predominance of proteins residing in the cytoplasm, plasma membrane and apoplast, among other cellular components (Supplementary Figure S8). The same comparison showed a predominance of proteins involved in transferase activity, nucleotide binding, and catalytic activity, among other molecular functions (Supplementary Figure S9).

3.5. Differential Expression of Cytoskeletal Related Proteins

Identifying proteins related to fiber development and maturation is essential for discovering the cellular network responsible for fiber and cell wall development. In this study, we identified multiple cytoskeletal-related proteins that were differentially regulated, including the actin binding proteins, profilin and actin-depolymerizing factor (ADF). Profilins have been found to be expressed during early cotton fiber development [53,54,55], and when overexpressed in transgenic tobacco cells, they produced an elongated cell with thicker and longer microfilament cables [54,55]. We found that an isoform of profilin (A0A0B0PF49) was downregulated during secondary wall deposition (25 dpa) when compared to an earlier fiber development stage (15 dpa), while actin depolymerizing factor 7 (A1XJ44) was found to be upregulated in 15 dpa when compared to 10 dpa and downregulated in 25 dpa fiber when compared to 15 dpa. Another cytoskeleton binding protein, known as annexin, was found to be downregulated in 15 dpa (Q69DC2, Q8W4Z7, A6MUT1) and 25 dpa fiber (Q69DC2, S5GI78, D2D2Z9, S5G619, M4MX81). Annexins are capable of binding to calcium and lipid membranes, allowing them to play a role in signaling networks, membrane trafficking and cytoskeletal interactions [56,57,58,59,60,61,62,63].

We also identified several beta and alpha-tubulin proteins that were differentially expressed, including one that was upregulated in 15 dpa fiber (A0A0B0MX93, tubulin beta-1 chain-like protein). Several others were upregulated in 25 dpa fiber, including alpha-tubulin 3 (Q8H6L9), alpha-tubulin 2 (Q8H6M0), tubulin beta-7 chain (A0A0B0PUW5), tubulin alpha-2 chain (A0A0B0NQH0), beta-tubulin 1 (A6MUS5) and tubulin beta-7 chain (A0A0B0NRH5). Alpha-tubulin 3 (Q8H6L9), alpha-tubulin 2 (Q8H6M0) and tubulin alpha-3/alpha-5 chain-like protein (A0A0B0PKT8) were downregulated in 15 dpa fiber. Other important differentially-regulated cytoskeletal proteins included an actin protein that was downregulated in 15 dpa fiber (A6MUT3); while four other actin proteins showed downregulation in 25 dpa fiber (Q7XZK1, A0A0B0NUJ5, A0A0B0MGZ2 and I1T3U6).

3.6. Differential Expression of Cell Wall-Related Proteins

Many proteins involved in cell wall synthesis and modification were found to be differentially expressed at various time points during the fiber development process. Interestingly, several expansin protein family members showed downregulation during fiber development. For instance, alpha-expansin (B2Z3V3, Q283Q7, Q8LKK3, Q8LKK2), expansin (Q7Y254,

O23952), alpha expansin 1 (A6YR48, I1T470, A6YR43, I1T469) and alpha expansin 2 (A6YR66 and A6YR63) were downregulated in 15 dpa fiber.

3.7. Differential Expression of Energy/Carbohydrate Metabolism Proteins

Many sucrose synthase isoforms were found to be differentially regulated. For instance, sucrose synthase proteins (I1T4T5, G9BRX6, I1T4T6, H6AC56, H6AC57 and G9BY14) were upregulated in 15 dpa fiber; while in 25 dpa fiber, sucrose synthase isoform C proteins were upregulated (G1JRK6, G1FNX7 and G1FNX4). Some sucrose synthase proteins were also downregulated. For example, sucrose synthase isoform B (G1JRK5), sucrose synthase isoform D (G1JRK7), sucrose synthase Sus1 (I1T4R3), sucrose synthase 2 (G9BRX6) and sucrose synthase (Q9ZRC4, I1T4T5, I1T4T6, H6AC56, H6AC57, G9BY14) were all downregulated in 25 dpa fiber. The large number of differentially-regulated energy/carbohydrate metabolism proteins suggested important roles of these proteins in fiber development, which is consistent with prior reports. Brill et al. showed that SusA/B/D isoforms are expressed highly during fiber elongation, falling off during secondary cell wall synthesis [64]. However, SusC is absent at both the transcript and protein levels in early fiber development, but highly expressed in later fiber development [64].

4. DISCUSSION

4.1. Comparison of Protein Extraction Methods for Recalcitrant Cotton Fiber

Protein extraction from recalcitrant plant tissue is highly challenging due to the abundance of interfering plant chemical compounds and extremely robust cell wall. Efficient and routine study of recalcitrant plant material has been limited due to the high amount of plant secondary metabolites, including phenols, flavonoids, stilbenes, terpenes, tannins and lignins, which negatively affect downstream proteomic efforts [65,66]. Secondary metabolites build up as soluble forms in the vacuoles and are more prevalent in adult mature tissues than in young etiolated tissues [66,67]. For proteome analysis of cotton fiber, an ideal protein extraction method should successfully disrupt the cell wall and reproducibly extract all of the proteins in the proteome, with efficient removal of non-protein contaminants [66]. In the present study, we evaluated plant protein extraction protocols, including phenol extraction and IEF buffer coupled with and without pressure cycling technology (PCT). A comparison of the protein extraction methods was done based on protein identification.

Considering protein identification, phenol plus PCT identified the largest amount of proteins (980), while phenol extraction came in second, identifying 934 proteins with two or more peptides in 25 dpa fiber. PCT used together with IEF buffer identified 884 proteins; without PCT, 845 proteins were identified in 25 dpa fiber. Our results suggested that to thoroughly study the cotton total proteome, the use of multiple extraction procedures collectively is favorable, because they provide a broader range of coverage of the total proteome. Combining the various protein extraction methods in this study, we identified 1446 proteins (Supplemental Table S1, Supplemental Tables S4 to S7 and Supplemental Figures S10 to S12) and 287 differentially-expressed proteins (Supplemental Table S3) in G. hirsutum. Our study provided a comprehensive proteome in G. hirsutum, including the proteome of 35 dpa fiber.

4.2. Proteome Studies on 35 dpa Fiber

Although many studies have been performed on different stages of cotton fiber development, the majority of these studies have primarily focused on early fiber stages due to the recalcitrant nature of maturing cotton fiber. Many studies cited the difficulty resulting from the tough secondary cell wall exterior and reduced protein content [7,46]. To the best of our knowledge, this is the first report to present proteome study data on 35 dpa fiber. We identified 785 proteins in the 35 dpa fiber proteome. Seventeen of these proteins were exclusively identified in 35 dpa fiber tissue. These proteins included three uncharacterized proteins (A0A0B0N1M7, A0A0B0MIW5, A0A0B0PNM1), glutamate—glyoxylate aminotransferase 2-like protein (A0A0B0NW95), peroxidase 51-like protein (A0A0B0PPW0), WRKY transcription factor 1-like protein (A0A0B0PEV3), NF-X1-type zinc finger NFXL1-like protein (A0A0B0NHR9), GDSL esterase/lipase CPRD49-like protein (A0A0B0N5L5), quinone oxidoreductase-like protein 2 (A0A0B0P6H8), Biotin carboxyl carrier protein subunit (A8RWF8), alcohol dehydrogenase class-P-like protein (A0A0B0NGV8), 60S acidic ribosomal P1 (A0A0B0PNW5), cell division cycle 48 (A0A0B0NEP6), UDP-glucuronic acid decarboxylase 1 (A0A0B0N8J8), polygalacturonase-inhibiting protein (Q6WMU5), lysosomal beta glucosidase (A0A0B0PUQ5) and beta-D-xylosidase 1-like protein (A0A0B0PLT0).

4.3. Proteome Differential Regulation during Cotton Fiber Development

In this study, we aimed to identify regulatory proteins that were expressed at specific developmental stages. We found 287 proteins that were differentially expressed at one or more of the three time points compared. However, the number of differentially-regulated proteins identified in this study may be inflated due to the presence of homologous proteins in the database, which can share peptides. It is difficult to distinguish whether all of the proteins sharing the same set of peptides are present in the sample or if only some of the proteins are actually present [68]. In future studies, identification of proteotypic peptides for quantitative proteomics would aid in overcoming this setback [68]. Many of the differentially-expressed proteins belonged to functional classes involved in cytoskeletal arrangement, energy/carbohydrate metabolism, stress responses and cell wall. The distribution of the differentially-regulated proteins indicates the cytoskeleton plays a critical role in cotton fiber development; as well as a strong role of energy/carbohydrate metabolism throughout fiber development. This result is plausible, because rapid cell elongation and fiber development require a large amount of energy, as well as cytoskeletal and cell wall dynamic rearrangement [8]. Furthermore, the distribution of the identified proteins shows that cotton fiber cells possess an increased level of redox-related activity at the 15 dpa fiber stage, likely due to H_2O_2 content increasing at 15 dpa and peaking at the 20 dpa stage of fiber development [8]. Altogether, protein functional analysis and regulation data indicate that fiber development involves an interplay between cytoskeletal, cell wall and metabolic proteins, creating a complex arrangement of events associated with key fiber developmental stages. Our observations were consistent with prior 2-DE gel-based studies on fiber development and comparison between the fuzzless-lintless mutant and wild-type.

4.4. Cytoskeletal Dynamics

Cytoskeletal architecture in cotton fibers plays an essential role in cotton fiber development. Plant cell morphology is largely determined by the highly dynamic actin cytoskeleton [55]. In this study, we found that cytoskeletal proteins were highly differentially regulated during fiber development, which included actin and actin binding proteins profilin and annexin. Actin was found to be downregulated in 15 dpa fiber and 25 dpa fiber when compared to earlier time points.

The activity of actin is further modulated by actin-modifying proteins, including profilins (PFN), which were found to be upregulated in 15 dpa cotton fiber in this study. Wang et al. found that expression of a profilin family member (GhPFN2) was significantly induced during the period of rapid fiber elongation and secondary wall synthesis [55,69]. Another important regulator of actin dynamics, annexin was found to be downregulated in 15 dpa and 25 dpa fiber. In cotton, it was shown that four cotton annexin proteins (AnxGh1: AAR13288, AnxGh2: AAB67993, AnxGhFx:FJ415173, AnxGhF: AAC33305) were present in higher amount in fibers of 10 dpa wild-type plants as compared to the fuzzless-lintless mutant [6,63]. Another actin regulating protein called actin depolymerizing factor (ADF) was found to be upregulated in 15 dpa and downregulated in

25 dpa. Interestingly, it was found that increased expression of profilin 2, as well as suppression of actin depolymerizing factor 1 led to inhibition of fiber elongation and promoted secondary wall formation [69,70,71].

Another important family of cytoskeletal proteins known as tubulins, specifically several beta and alpha-tubulin isoforms, were differentially expressed. The bulk of these proteins were upregulated in 25 dpa fiber. During fiber elongation, a minimum of nine β-tubulin genes are preferentially or differentially expressed [72]. A study by Pu et al. in cotton fiber suggested that transcription factors, such as GhMYB109, control key regulatory processes by modulating microtubules by influencing the expression of GhTUB1 and GhACT1 genes, thereby effecting subsequent downstream cellular effects related to the cytoskeleton [72,73]. Therefore, the differential regulation of actin and tubulin variants along with actin-regulating proteins may be part of a complex regulatory network of proteins controlling fiber development.

4.5. Cell Wall Dynamics

Cotton fiber provides a good model for studying the regulation of primary and secondary cell wall development. During rapid fiber elongation, it is known that genes encoding wall-loosening expansin proteins [74,75] are expressed and are associated with quantitative trait loci linked with fiber length [72,76,77]. In this study, expansins were also found to be differentially regulated. The expansin family members were downregulated in 15 dpa with expression patterns consistent with previous studies. The expansin gene plays a very important role during cell wall extension [78,79]. Ruan et al. found that the transcript of the fiber-specific wall loosening gene, GhEXP1, was high at the early phase of elongation (6 to 8 dpa), but substantially reduced to untraceable levels at 20 dpa, suggesting that at this point, the primary cell wall of elongating fiber has become very rigid, elongation has stopped because of the loss of higher turgor as a result of downregulation of transporter genes and reopening of plasmodesmata and fiber cells have shifted their energy to cellulose synthesis [79].

5. CONCLUSIONS

In this study, protocols for protein extraction from mature cotton fiber tissue were developed. Combining different protein extraction protocols and LC-MS/MS technology, we have established a comprehensive cotton fiber proteome in Gossypium hirsutum, including the 35 dpa fiber proteome, which has not been examined in prior reports to the best of our knowledge. Proteins with diverse functions at specific developmental stages of cotton fiber were identified by quantitative comparison at three representative developmental stages. The results provide novel insight into cotton fiber development and regulation in addition to identifying efficient methods of protein extraction for cotton fiber proteomics.

REFERENCES

1. Chen, Z.J.; Scheffler, B.E.; Dennis, E.; Triplett, B.A.; Zhang, T.; Guo, W.; Chen, X.; Stelly, D.M.; Rabinowicz, P.D.; Town, C.D.; et al. Toward sequencing cotton (Gossypium) genomes. Plant Physiol. **2007**, 145, 1303–1310.
2. Zhang, B.; Yang, Y.W.; Zhang, Y.; Liu, J.Y. A high-confidence reference dataset of differentially expressed proteins in elongating cotton fiber cells. Proteomics **2013**, 13, 1159–1163.
3. Kim, H.J.; Triplett, B.A. Cotton fiber growth in planta and in vitro. Models for plant cell elongation and cell wall biogenesis. Plant Physiol. **2001**, 127, 1361–1366.
4. Singh, B.; Avci, U.; Eichler Inwood, S.E.; Grimson, M.J.; Landgraf, J.; Mohnen, D.; Sorensen, I.; Wilkerson, C.G.; Willats, W.G.; Haigler, C.H. A specialized outer layer of the primary cell wall joins elongating cotton fibers into tissue-like bundles. Plant Physiol. **2009**, 150, 684–699.
5. Zhang, B.; Liu, J.Y. Mass spectrometric identification of in vivo phosphorylation sites of differentially expressed proteins in elongating cotton fiber cells. PLoS ONE **2013**, 8, e58758.
6. Pang, C.Y.; Wang, H.; Pang, Y.; Xu, C.; Jiao, Y.; Qin, Y.M.; Western, T.L.; Yu, S.X.; Zhu, Y.X. Comparative proteomics indicates that biosynthesis of pectic precursors is important for cotton fiber and arabidopsis root hair elongation. Mol. Cell Proteomics **2010**, 9, 2019–2033.
7. Yao, Y.; Yang, Y.W.; Liu, J.Y. An efficient protein preparation for proteomic analysis of developing cotton fibers by 2-de. Electrophoresis **2006**, 27, 4559–4569.
8. Yang, Y.W.; Bian, S.M.; Yao, Y.; Liu, J.Y. Comparative proteomic analysis provides new insights into the fiber elongating process in cotton. J. Proteome Res. **2008**, 7, 4623–4637.
9. Haigler, C.H.; Betancur, L.; Stiff, M.R.; Tuttle, J.R. Cotton fiber: A powerful single-cell model for cell wall and cellulose research. Front. Plant Sci. **2012**, 3, 104.

10. Li, Y.J.; Wang, F.X.; Wang, Y.Q.; Liu, Y.C.; Zhang, X.Y.; Sun, Y.Q.; Sun, J. Identification of the proteins in green cotton fiber using a proteomics-based approach. Biotechnol. Lett. **2013**, 35, 1519–1523.
11. Li, Y.J.; Zhang, X.Y.; Wang, F.X.; Yang, C.L.; Liu, F.; Xia, G.X.; Sun, J. A comparative proteomic analysis provides insights into pigment biosynthesis in brown color fiber. J. Proteomics **2013**, 78, 374–388.
12. Arpat, A.B.; Waugh, M.; Sullivan, J.P.; Gonzales, M.; Frisch, D.; Main, D.; Wood, T.; Leslie, A.; Wing, R.A.; Wilkins, T.A. Functional genomics of cell elongation in developing cotton fibers. Plant Mol. Biol. **2004**, 54, 911–929.
13. Zhao, P.M.; Wang, L.L.; Han, L.B.; Wang, J.; Yao, Y.; Wang, H.Y.; Du, X.M.; Luo, Y.M.; Xia, G.X. Proteomic identification of differentially expressed proteins in the ligon lintless mutant of upland cotton (Gossypium hirsutum L.). J. Proteome Res. **2010**, 9, 1076–1087.
14. Lee, J.J.; Hassan, O.S.; Gao, W.; Wei, N.E.; Kohel, R.J.; Chen, X.Y.; Payton, P.; Sze, S.H.; Stelly, D.M.; Chen, Z.J. Developmental and gene expression analyses of a cotton naked seed mutant. Planta **2006**, 223, 418–432.
15. Wu, Y.; Machado, A.C.; White, R.G.; Llewellyn, D.J.; Dennis, E.S. Expression profiling identifies genes expressed early during lint fibre initiation in cotton. Plant Cell Physiol. **2006**, 47, 107–127.
16. Liang, W.; Fang, L.; Xiang, D.; Hu, Y.; Feng, H.; Chang, L.; Zhang, T. Transcriptome analysis of short fiber mutant ligon lintless-1 (li1) reveals critical genes and key pathways in cotton fiber elongation and leaf development. PLoS ONE **2015**, 10, e0143503.
17. Yao, Y.; Zhang, B.; Dong, C.J.; Du, Y.; Jiang, L.; Liu, J.Y. Comparative proteomic and biochemical analyses reveal different molecular events occurring in the process of fiber initiation between wild-type allotetraploid cotton and its fuzzless-lintless mutant. PLoS ONE **2015**, 10, e0117049.
18. Wang, Q.Q.; Liu, F.; Chen, X.S.; Ma, X.J.; Zeng, H.Q; Yang, Z.M. Transcriptome profiling of early developing cotton fiber by deep-sequencing reveals significantly differential expression of genes in a fuzzless/lintless mutant. Genomics **2010**, 96, 369–376.
19. Padmalatha, K.V.; Patil, D.P.; Kumar, K.; Dhandapani, G.; Kanakachari, M.; Phanindra, M.L.; Kumar, S.; Mohan, T.C.; Jain, N.; Prakash, A.H.; et al. Functional genomics of fuzzless-lintless mutant of Gossypium hirsutum L. Cv. Mcu5 reveal key genes and pathways involved in cotton fibre initiation and elongation. BMC Genomics **2012**, 13, 624.

20. Liu, K.; Han, M.; Zhang, C.; Yao, L.; Sun, J.; Zhang, T. Comparative proteomic analysis reveals the mechanisms governing cotton fiber differentiation and initiation. J. Proteomics **2012**, 75, 845–856.
21. Liu, K.; Sun, J.; Yao, L.; Yuan, Y. Transcriptome analysis reveals critical genes and key pathways for early cotton fiber elongation in ligon lintless-1 mutant. Genomics **2012**, 100, 42–50.
22. Du, S.J.; Dong, C.J.; Zhang, B.; Lai, T.F.; Du, X.M.; Liu, J.Y. Comparative proteomic analysis reveals differentially expressed proteins correlated with fuzz fiber initiation in diploid cotton (Gossypium arboreum L.). J. Proteomics **2013**, 82, 113–129.
23. Graves, D.A.; Stewart, J.M. Analysis of the protein constituency of developing cotton fibres. J. Exp. Bot. **1988**, 39, 59–69.
24. Hurkman, W.J.; Tanaka, C.K. Solubilization of plant membrane proteins for analysis by two-dimensional gel electrophoresis. Plant Physiol. **1986**, 81, 802–806.
25. Turley, R.B.; Ferguson, D.L. Changes of ovule proteins during early fiber development in a normal and a fiberless line of cotton (Gossypium hirsutum L.). J. Plant. Physiol. **1996**, 149, 695–702.
26. Tao, F.; Lawrence, N.P.; Miller, W.W.; Li, C.; Tuzmen, P.; Behnke, J.; Nakhai, B.; Kakita, A.; Christian, T.; Reed, D.; et al. Biological sample preparation system using Pressure Cycling Technology (PCT). In Advances in High Pressure Bioscience and Biotechnology II; Winter, R., Ed.; Springer: Dortmund, Germany, 2003; pp. 413–417.
27. Tao, F.; Schumacher, R.T.; Lawrence, N.P.; Gross, V.; Smejkal, G.; Lazarev, A.; Han, R. Extraction of Biomolecular Complexes Assisted by Alternating Hydrostatic Pressure. United States Patent US 2009/0286301, 9 November 2009. Available online: http://www.freepatentsonline.com/y2009/0286301.html (accessed on 6 July 2015).
28. Freeman, E.; Ivanov, A.R. Proteomics under pressure: Development of essential sample preparation techniques in proteomics using ultrahigh hydrostatic pressure. J. Proteome Res. **2011**, 10, 5536–5546.
29. Smejkal, G.B.; Robinson, M.H.; Lawrence, N.P.; Tao, F.; Saravis, C.A.; Schumacher, R.T. Increased protein yields from Escherichia coli using pressure-cycling technology. J. Biomol. Tech. **2006**, 17, 173–175.
30. Ringham, H.; Bell, R.L.; Smejkal, G.B.; Behnke, J.; Witzmann, F.A. Application of pressure cycling technology to tissue sample preparation for 2-DE. Electrophoresis **2007**, 28, 1022–1024.

31. Smejkal, G.; Witzmann, F.; Ringham, H.; Small, D.; Chase, S.; Behnke, J.; Ting, E. Sample preparation for two-dimensional gel electrophoresis using pressure cycling technology. Anal Biochem. **2007**, 363, 309–311.
32. Szabo, Z.; Guttman, A.; Karger, B.L. Rapid release of n-linked glycans from glycoproteins by pressure-cycling technology. Anal. Chem. **2010**, 82, 2588–2593.
33. Meersman, F.; Dobson, C.M.; Heremans, K. Protein unfolding, amyloid fibril formation and configurational energy landscapes under high pressure conditions. Chem. Soc. Rev. **2006**, 35, 908–917.
34. Guttman, A.; Chen, F.T.; Evangelista, R.A.; Cooke, N. High-resolution capillary gel electrophoresis of reducing oligosaccharides labeled with 1-aminopyrene-3,6,8-trisulfonate. Anal. Biochem. **1996**, 233, 234–242.
35. Reddy, K.R.; Hodges, H.; Read, J.; McKinion, J.; Baker, J.; Tarpley, L.; Reddy, V. Soil-plant-atmosphere-research (spar) facility: A tool for plant research and modeling. Biotronics **2001**, 30, 27–50.
36. Hewitt, E.J. Sand and Water Culture Methods used in the Study of Plant Nutrition. Technical Communication 22, 1952; Commonwealth Agriculture Bureaux, Farham Royal: Bucks, UK.
37. Chitteti, B.R.; Peng, Z. Proteome and phosphoproteome differential expression under salinity stress in rice (Oryza sativa) roots. J. Proteome Res. **2007**, 6, 1718–1727.
38. Tan, F.; Zhang, K.; Mujahid, H.; Verma, D.P.; Peng, Z. Differential histone modification and protein expression associated with cell wall removal and regeneration in rice (Oryza sativa). J. Proteome Res. **2011**, 10, 551–563.
39. Mujahid, H.; Tan, F.; Zhang, J.; Nallamilli, B.R.; Pendarvis, K.; Peng, Z. Nuclear proteome response to cell wall removal in rice (Oryza sativa). Proteome Sci. **2013**, 11, 26.
40. Shah, P.; Nanduri, B.; Swiatlo, E.; Ma, Y.; Pendarvis, K. Polyamine biosynthesis and transport mechanisms are crucial for fitness and pathogenesis of streptococcus pneumoniae. Microbiology **2011**, 157, 504–515.
41. Tan, F.; Li, G.; Chitteti, B.R.; Peng, Z. Proteome and phosphoproteome analysis of chromatin associated proteins in rice (Oryza sativa). Proteomics **2007**, 7, 4511–4527.
42. Chitteti, B.R.; Tan, F.; Mujahid, H.; Magee, B.G.; Bridges, S.M.; Peng, Z. Comparative analysis of proteome differential regulation during cell dedifferentiation in arabidopsis. Proteomics **2008**, 8, 4303–4316.

43. Kessner, D.; Chambers, M.; Burke, R.; Agus, D.; Mallick, P. Proteowizard: Open source software for rapid proteomics tools development. Bioinformatics **2008**, 24, 2534–2536.
44. Craig, R.; Beavis, R.C. Tandem: Matching proteins with tandem mass spectra. Bioinformatics **2004**, 20, 1466–1467.
45. Protein Grouping. Available online: https://github.com/jsreddy82/ProteinGrouping.git (accessed on 3 May 2015).
46. Wan, C.Y.; Wilkins, T.A. A modified hot borate method significantly enhances the yield of high-quality RNA from cotton (Gossypium hirsutum L.). Anal. Biochem. **1994**, 223, 7–12.
47. Oliveros, J.C. (2007–2015) Venny. An interactive tool for comparing lists with Venn's diagrams. Available online: http://bioinfogp.cnb.csic.es/tools/venny/index.html (accessed on 20 May 2015).
48. Davis, M.J.; Sehgal, M.S.; Ragan, M.A. Automatic, context-specific generation of gene ontology slims. BMC Bioinform. **2010**, 11, 498.
49. Al-Ghazi, Y.; Bourot, S.; Arioli, T.; Dennis, E.S.; Llewellyn, D.J. Transcript profiling during fiber development identifies pathways in secondary metabolism and cell wall structure that may contribute to cotton fiber quality. Plant Cell Physiol. **2009**, 50, 1364–1381.
50. Payne, A.; Schmidt, T.B.; Nanduri, B.; Pendarvis, K.; Pittman, J.R.; Thornton, J.A.; Grissett, J.; Donaldson, J.R. Proteomic analysis of the response of listeria monocytogenes to bile salts under anaerobic conditions. J. Med. Microbiol. **2013**, 62, 25–35.
51. McCarthy, F.M.; Wang, N.; Magee, G.B.; Nanduri, B.; Lawrence, M.L.; Camon, E.B.; Barrell, D.G.; Hill, D.P.; Dolan, M.E.; Williams, W.P.; et al. Agbase: A functional genomics resource for agriculture. BMC Genomics **2006**, 7, 229.
52. McCarthy, F.M.; Bridges, S.M.; Wang, N.; Magee, G.B.; Williams, W.P.; Luthe, D.S.; Burgess, S.C. Agbase: A unified resource for functional analysis in agriculture. Nucleic Acids Res. **2007**, 35, D599–D603.
53. Ji, S.J.; Lu, Y.C.; Feng, J.X.; Wei, G.; Li, J.; Shi, Y.H.; Fu, Q.; Liu, D.; Luo, J.C.; Zhu, Y.X. Isolation and analyses of genes preferentially expressed during early cotton fiber development by subtractive PCR and CDNA array. Nucleic Acids Res. **2003**, 31, 2534–2543.
54. Wang, H.Y.; Yu, Y.; Chen, Z.L.; Xia, G.X. Functional characterization of Gossypium hirsutum profilin 1 gene (ghpfn1) in tobacco suspension

cells. Characterization of in vivo functions of a cotton profilin gene. Planta **2005**, 222, 594–603.

55. Bao, Y.; Hu, G.; Flagel, L.E.; Salmon, A.; Bezanilla, M.; Paterson, A.H.; Wang, Z.; Wendel, J.F. Parallel up-regulation of the profilin gene family following independent domestication of diploid and allopolyploid cotton (Gossypium). Proc. Natl. Acad. Sci. USA **2011**, 108, 21152–21157.
56. Diaz-Munoz, M.; Hamilton, S.L.; Kaetzel, M.A.; Hazarika, P.; Dedman, J.R. Modulation of Ca^{2+} release channel activity from sarcoplasmic reticulum by annexin vi (67-kda calcimedin). J. Biol. Chem. **1990**, 265, 15894–15899.
57. Alvarez-Martinez, M.T.; Porte, F.; Liautard, J.P.; Sri Widada, J. Effects of profilin-annexin I association on some properties of both profilin and annexin I: Modification of the inhibitory activity of profilin on actin polymerization and inhibition of the self-association of annexin I and its interactions with liposomes. Biochim. Biophys. Acta **1997**, 1339, 331–340.
58. Carroll, A.D.; Moyen, C.; van Kesteren, P.; Tooke, F.; Battey, N.H.; Brownlee, C. Ca^{2+}, annexins, and GTP modulate exocytosis from maize root cap protoplasts. Plant Cell **1998**, 10, 1267–1276.
59. Lafont, F.; Lecat, S.; Verkade, P.; Simons, K. Annexin xiiib associates with lipid microdomains to function in apical delivery. J. Cell Biol. **1998**, 142, 1413–1427.
60. Reddy, V.S.; Reddy, A.S. Proteomics of calcium-signaling components in plants. Phytochemistry **2004**, 65, 1745–1776.
61. Konopka-Postupolska, D. Annexins: Putative linkers in dynamic membrane-cytoskeleton interactions in plant cells. Protoplasma **2007**, 230, 203–215.
62. Morel, E.; Gruenberg, J. Annexin a2 binding to endosomes and functions in endosomal transport are regulated by tyrosine 23 phosphorylation. J. Biol. Chem. **2009**, 284, 1604–1611.
63. Huang, Y.; Wang, J.; Zhang, L.; Zuo, K. A cotton annexin protein anxgb6 regulates fiber elongation through its interaction with actin 1. PLoS ONE **2013**, 8, e66160.
64. Brill, E.; van Thournout, M.; White, R.G.; Llewellyn, D.; Campbell, P.M.; Engelen, S.; Ruan, Y.L.; Arioli, T.; Furbank, R.T. A novel isoform of sucrose synthase is targeted to the cell wall during secondary cell wall synthesis in cotton fiber. Plant Physiol. **2011**, 157, 40–54.

65. Rhodes, M.J. Physiological roles for secondary metabolites in plants: Some progress, many outstanding problems. Plant Mol. Biol. **1994**, 24, 1–20.
66. Sebastiana, M.; Figueiredo, A.; Monteiro, F.; Martins, J.; Franco, C.; Coelho, A.V.; Vaz, F.; Simoes, T.; Penque, D.; Pais, M.S.; et al. A possible approach for gel-based proteomic studies in recalcitrant woody plants. Springerplus **2013**, 2, 210.
67. Granier, F. Extraction of plant proteins for two-dimensional electrophoresis. Electrophoresis **1988**, 9, 712–718.
68. Huang, T.; Wang, J.; Yu, W.; He, Z. Protein inference: A review. Brief Bioinform. **2012**, 13, 586–614.
69. Wang, J.; Wang, H.Y.; Zhao, P.M.; Han, L.B.; Jiao, G.L.; Zheng, Y.Y.; Huang, S.J.; Xia, G.X. Overexpression of a profilin (ghpfn2) promotes the progression of developmental phases in cotton fibers. Plant Cell Physiol. **2010**, 51, 1276–1290.
70. Wang, H.Y.; Wang, J.; Gao, P.; Jiao, G.L.; Zhao, P.M.; Li, Y.; Wang, G.L.; Xia, G.X. Down-regulation of ghadf1 gene expression affects cotton fibre properties. Plant Biotechnol. J. **2009**, 7, 13–23.
71. Lee, J.; Burns, T.H.; Light, G.; Sun, Y.; Fokar, M.; Kasukabe, Y.; Fujisawa, K.; Maekawa, Y.; Allen, R.D. Xyloglucan endotransglycosylase/hydrolase genes in cotton and their role in fiber elongation. Planta **2010**, 232, 1191–1205.
72. Stiff, M.R.; Haigler, C.H. Recent advances in cotton fiber development. In Flowering and Fruiting in Cotton; Oosterhuis, D.M., Cothren, J.T., Eds.; The Cotton Foundation: Cordova, TN, USA, 2012; Volume 8, pp. 163–192.
73. Pu, L.; Li, Q.; Fan, X.; Yang, W.; Xue, Y. The r2r3 MYB transcription factor ghmyb109 is required for cotton fiber development. Genetics **2008**, 180, 811–820.
74. Sampedro, J.; Cosgrove, D.J. The expansin superfamily. Genome Biol. **2005**, 6, 242.
75. Cosgrove, D.J. Growth of the plant cell wall. Nat. Rev. Mol. Cell Biol. **2005**, 6, 850–861.
76. Harmer, S.E.; Orford, S.J.; Timmis, J.N. Characterisation of six alpha-expansin genes in Gossypium hirsutum (upland cotton). Mol. Genet. Genomics **2002**, 268, 1–9.
77. An, C.; Saha, S.; Jenkins, J.N.; Scheffler, B.E.; Wilkins, T.A.; Stelly, D.M. Transcriptome profiling, sequence characterization, and snp-based

chromosomal assignment of the expansin genes in cotton. Mol. Genet. Genomics **2007**, 278, 539–553.
78. Cosgrove, D.J. Relaxation in a high-stress environment: The molecular bases of extensible cell walls and cell enlargement. Plant Cell **1997**, 9, 1031–1041.
79. Ruan, Y.L.; Llewellyn, D.J.; Furbank, R.T. The control of single-celled cotton fiber elongation by developmentally reversible gating of plasmodesmata and coordinated expression of sucrose and K^+ transporters and expansin. Plant Cell **2001**, 13, 47–60.

7

CHAPTER

A PPO PROMOTER FROM BETALAIN-PRODUCING RED SWISS CHARD, DIRECTS PETIOLE- AND ROOT-PREFERENTIAL EXPRESSION OF FOREIGN GENE IN ANTHOCYANINS-PRODUCING PLANTS

Zhi-Hai Yu, Ya-Nan Han and Xing-Guo Xiao *

State Key Laboratory of Plant Physiology and Biochemistry, College of Biological Sciences, China Agricultural University, Beijing 100193, China

ABSTRACT

A 1670 bp 5′-flanking region of the polyphenol oxidase (PPO) gene was isolated from red Swiss chard, a betalain-producing plant. This region, named promoter BvcPPOP, and its 5′-truncated versions were fused with the GUS gene and introduced into Arabidopsis, an anthocyanins-producing plant. GUS histochemical staining and quantitative analysis of transgenic plants at the vegetative and reproductive stages showed that BvcPPOP could direct GUS gene expression in vegetative organs with root- and petiole-preference, but not in reproductive organs including inflorescences shoot, inflorescences leaf, flower, pod and seed. This promoter was regulated by developmental stages in its driving strength. but not in expression pattern. It was also regulated by the abiotic stressors tested, positively by salicylic acid (SA) and methyl jasmonate (MeJA) but negatively by abscisic acid (ABA), gibberellin (GA), NaCl and OH^-. Its four 5′-truncated versions varied in the driving strength, but not obviously in expression pattern, and even the shortest version (−225 to +22) retained the root- and petiole- preference. This promoter is, to our knowledge, the first PPO promoter cloned and functionally elucidated from the betalain-producing plant, and thus provides not only a useful tool for expressing gene(s) of agricultural interest in vegetative organs, but also a clue to clarify the function of metabolism-specific PPO in betalain biosynthesis.

Keywords: anthocyanin-producing plant; betalain-producing plant; Beta vulgaris subsp. cicla; development stage; petiole-preference; polyphenol oxidase (PPO); promoter; root-preferential expression; vegetative organ

1. INTRODUCTION

Polyphenol oxidases (PPOs) are type-3 copper enzymes that can catalyze the o-hydroxylation of monophenols (monophenolase activity, EC 1.14.18.1) and oxidation of o-diphenols to o-quinones (diphenolase activity, catechol oxidase activity, EC 1.10.3.1). They are widely distributed in plants, animals and microbes. In plants, PPOs can be classified into two groups based on their physiological roles: Group 1: PPOs engaged in adaptation and defense against biotic and abiotic stresses; Group 2: PPOs involved in some specific secondary metabolisms [1,2]. For Group 1 PPOs, their physiological roles have been linked to PPO genes [3,4,5,6,7,8] and a grand body of PPO genes have their promoters cloned and functionally analyzed, such as the PPO promoters in tomato [9,10], eggplant [11], dandelion [12], pineapple [13], rice [14], artichoke [15] and rubber tree [16] etc. However, for Group 2 PPOs, their involvement in specific secondary metabolism have been proposed based mainly on biochemical analysis, and only PPOs involved in the biosynthesis of aurones [17,18], lignans [19] and in tyrosine metabolism [20] have been approved by gene functional analysis.

PPO has long been proposed to be involved in the biosynthesis of betalains by sequential hydroxylation of l-tyrosine to form l-DOPA and oxidation of the DOPA to produce dopa-quinone [21,22,23] and this proposition has been widely adopted [24,25,26,27,28,29]. However, the molecular mechanisms in support of this hypothesis, to our knowledge, remain unclear. Joy and colleagues [30] cloned and sequence-characterized two PPO genes from Phytolacca Americana, but did not report their function. Kaneda and co-workers [31] isolated five PPO genes from Portulaca oleracea, and showed that two of them, PoPPO2 and PoPPO5 could catalyze hydroxylation of octylphenol, nonylphenol and 17β-estradiol. Whether these two PPOs and the rest can mediate conversion of l-tyrosine remains unknown. Our team purified and characterized a PPO from red Swiss chard (Beta vulgaris subspecies cicla), a betalain-rich vegetable [32] and cloned a PPO gene [33]. In this study, we report isolation, sequence and functional analysis of this PPO gene's promoter, together with its 5′-deleted versions. This promoter, named BvcPPOP, directs GUS gene expression in the vegetative organs of the transgenic anthocyanin-producing plant, Arabidopsis, with a root- and petiole-preferential pattern. The 5′-deleted versions did as well, but at different levels. The information reported here may help to understand whether the PPO is engaged in betalain biosynthesis, at least in red Swiss chard, and the promoter and its 5′-deleted versions disclosed here may provide useful tools for crop improvement via biotechnology route.

```
-1648   GCAAATGGTC TATATCCGAC ATCATTTACC TTGAGGGTAC AATGCTAATA AGACAAAACA
-1588   TTTGTTAATT GAGATAAACT CGATCAAGCT CGACTTGAGA AAAACCCCAT TTTATGAAAA
              F2▲
-1528   CTTGGTTAAA ATAAATTTGA CTAGACTAAG GTTGACTCAA CTCCCACTCA ACTTAAAATA
-1468   CTAGCGAGTT AAATTAATCT ACTCCAAACT TAAATTCAAT TCGTATCATT TATACTTGAT
-1408   TTTTTTTATA AAAAAGATAA AATCAATATC CGTTGCATAA GTTGTGAGAT TGTTTGGTAA
-1348   TCCAACTTTT TAAGGGGCGT AAAATGAACT CTTATCAAAT ACAAGGGTGA ATAAAGAAGG
-1288   CCAACTTTAA TTCTTCCATT TACAAGGACT TTACCAACTA ATTAGTTTAG TTAGTTGATA
-1228   CGTAGCGAGT AGTAGCAGTA CAAGTTGTAA AAGAAAGAGA GAGAATTTCT AGGAGGAAAG
           F3▲
-1168   AATCCCTCAT TGTTTGAGCC ATCAGTGAAC AGTTTCTTGT CATTTGAAAT AAGATTATTA
-1108   TAATAATCTC ACAATCTTAA AGAAAGAGAA ATATAATACT TCCTCCGTTT CTAAATAAGT
-1048   GCAACATTTC TTGTTTGCAC GTTTGCCAAT GCATTATTTT AACCATTAAT ATCTTCAATT
-988    ATCAATAAGT AAAAATTATA AAAAGTTGAT ATTAAAAAAA TTCACATTCA GACGAATCTA
-928    ACAAGATCCC ACATGACTAT GTTTTTCCTA AAATGTAAAT CACGAAAGCT AATTAAAGTT
-868    TAAACTGTGA ATAGTAAACA CTATGCAAAT GTTGCACTTG TTTTGAAACG GAGGAAGTAT
-808    AATACAGTAT GTGTATGTCT TGAGTGTAAC ATGTTACATT ATAATTTATT GTGTAACAAA
          F4▲
-748    ACTTAAATGT AATATAATGA CACATAATAA TATAAAATCC AACAAAATAA TATAAATTTC
-688    ATGTTATTTT GTCATTAATT TTATGTCATC TAAGTTTTTT TTTATAAAAA ATTTATAATT
-628    TATGGGTTGC TAAAATCAAA AAAAAATTAT ACTTCCTCTC ACTCAAAAAT ACTTTCTCCG
-568    TTTTTTAATA AGTGACAGAG AAAAAATGGA GAAATACAAA GGAGCCTAGA GAGAAAAAGT
-508    TGGGACTTAC CACTTGTTCA ACCAATCAAA TTTGAATATT GGATAAAGTA GGTAAGTGAA
-448    ATAATTAAAT GACGGCAAAG ATGTAACAAA TGAAAAGTAT TTTTTACTAC TTTGAGCTAT
-388    GTGTCACTTA TTGTGACATT TTCCTAAAAG AAAAACATAT CACTTATTAA AAAACCGAGA
-328    AAATATAAGA CATTTACATT TTACGAAAAT TTGCTTTTAT AAGTCACCAT GTGCTGGATC
-268    TGTGTCCATA TTCTATTATC TATATGCAAT TTGATCAATA GTTTACTAGT CGTAAAGTAG
                                                   F5▲
-208    CAGTTCTCTC AATAATACAT AAAGCAGCCA ATGGAGAACA CCTAATAGTC ACATTAGTTT
-148    AATGCTTATC ACGTGGACCC TGTCGGCTGT TTGAATTTTC AAATTTTTTT ATAGTATTTG
-88     TAGCAAGAAA ATTCCCTCAC ATTGATTTTT GGGGTTAGAC CAACAACCCC TGAAGTTACT
-28     ATAAAAGGAA ACGTAATCTA TCCACAATAt tccaccaaat atataaaaga
        TATAbox               Possible TSS                 R▲
```

Figure 1. Nucleotide sequence of the 5′-flanking region of PPO gene. Nucleotides are numbered on the left with the possible transcription start site which is designated as +1 and white-boxed. The TATA-box is underlined. The deletion positions are indicated with a black triangle behind the short name of forward (F2 to F5) and reverse (R) primers.

2. RESULTS

2.1. Cloning and Sequence Analysis of the BvcPPOP Promoter

A 1670-bp promoter region was successfully amplified using the PCR method and deposited in GenBank (accession No. KT285044). In this region, a transcription start site (TSS), a TATA-box which is 22 bp upstream of the TSS (Figure 1) and a large number of other cis-acting elements (Table S1) were predicted based on New PLACE [34] and PlantCARE [35] analysis.

2.2. BvcPPOP Directed Petiole- and Root-Preferential Expression of GUS Gene at the Vegetative Stage of Transgenic Arabidopsis

As shown in F1 of Figure 2, the GUS gene driven by the full-length BvcPPOP promoter (F1) was expressed very strongly in the root, hypocotyl, petiole and leaf base of transgenic plantlets aged 5 to 10 days. As plantlets grew (15-day-old), the GUS blue became lighter in some fibrous roots (F1 in Figure 2) except the root-tip (Figure S1), whereas the GUS staining in the petiole, hypocotyl and leaf base remained visibly unchanged, with much stronger staining in root and petiole than in leaf blade. Measured at 15-day-old, the GUS enzymatic activities in the root and petiole were 13-fold and 10-fold higher than that in the leaf, respectively (F1 in Figure 3A). In contrast, the GUS activity of the leaf in 35S-GUS transgenic plants was higher than that of root and petiole (35S in Figure 3A). At the whole plant level, the GUS activity in F1-GUS plants was significantly lower than that in 35S-GUS plants, being approximately 30% (5-day-old), 40% (10-day-old), and 25% (15-day-old) of that in 35S-GUS plants (Figure 3B). However, this difference was negligible in petioles (Figure 3A).

2.3. BvcPPOP-Directed GUS Expression Is Regulated by Developmental Stages

The expression pattern of the GUS gene driven by BvcPPOP changed when the plants entered the reproductive stage. At flowering and after anthesis, the GUS blue in the hypocotyl disappeared completely and that in petiole and the main veins of the rosette leaf faded progressively, although the GUS-staining in root remained strong ("after anthesis" in Figure 2, and GUS activity in Figure S2). In the reproductive organs of broad-sense, such as inflorescence shoot, inflorescence leaf, flower, pod and seed, no GUS blue was visible.

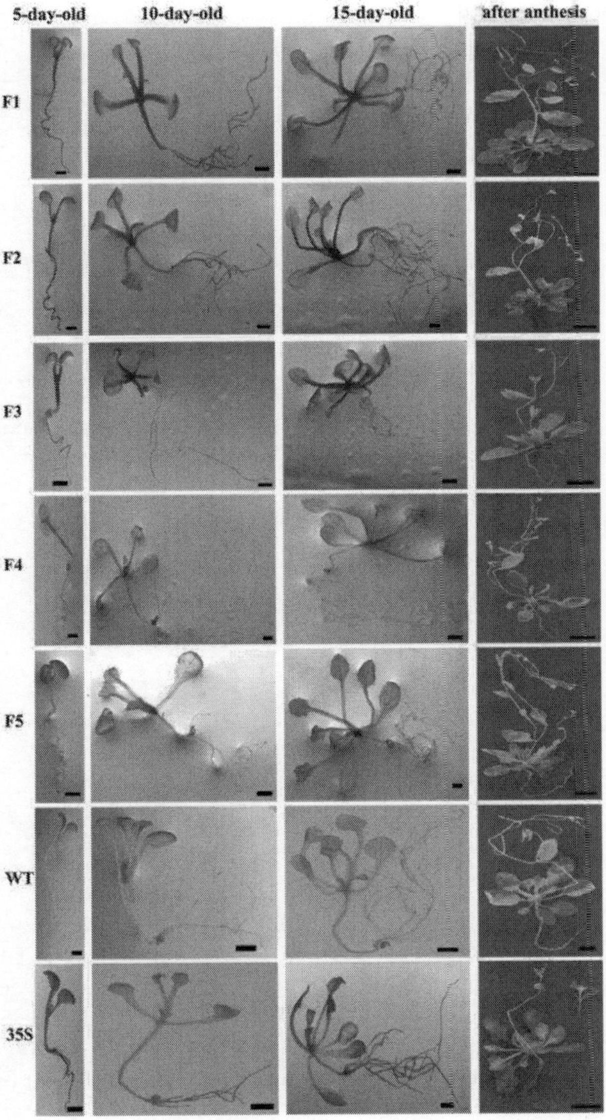

Figure 2. GUS staining of T_3 Arabidopsis. Bar = 1 mm, Bar = 1 cm ("after anthesis"). For each construct, three independent lines (more than 20 plantlets of each line) were GUS-stained; representative plants are shown.

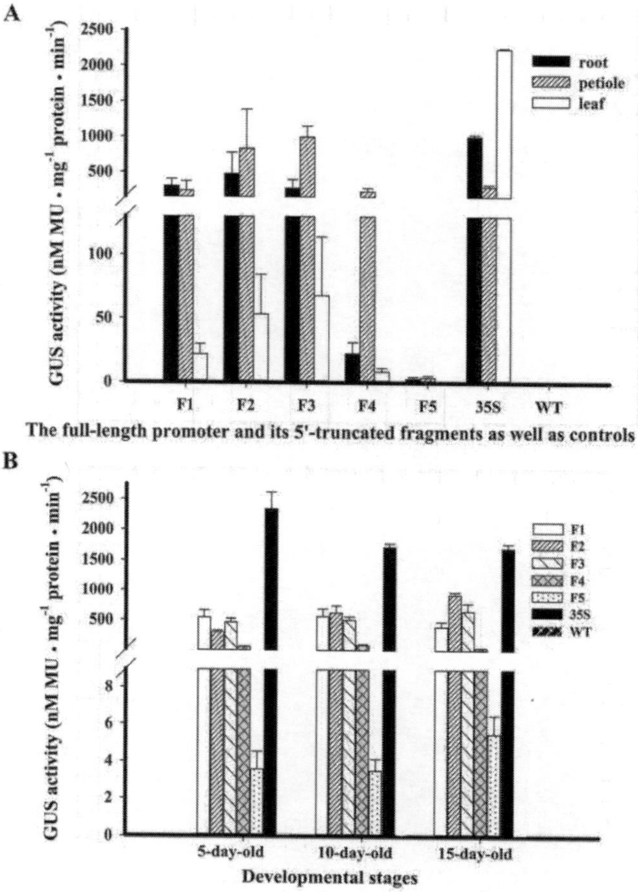

Figure 3. The GUS activities of T_3 Arabidopsis. Data are shown as mean values of three independent lines of each construct ± SD. (**A**) The GUS activity measured in root, petiole and leaf of 15-day-old different construct transgenic seedlings; (**B**) GUS activity driven by serial 5′-truncated BvcPPOPs and CaMV 35S as well as WT control in different developmental stages. GUS activity of single line of each construct was present in Figure S3.

2.4. BvcPPOP-Directed GUS Expression Is Differently Regulated by Abiotic Stressors

BvcPPOP-directed GUS expression was down-regulated by abiotic stressors tested in Set 1. NaCl (120 mM) treatment of 3-day-old F1-GUS transgenic seedlings for one week, resulted in GUS activity decline by around 79%, compared with that untreated, and the ABA (2 μM) or OH^- (pH 8.0) decreased about 70%, 60% of the GUS activity, respectively, while mannitol

(200 mM) resulted in only an approximately 19% decrement of the activity (Figure 4A).

The regulation of BvcPPOP-directed GUS expression was varied in Set 2 stressors. SA (1 mM) and MeJA (100 μM) root-treatment of 10-day-old plantlets for 24 h increased GUS activity by 21% and 84%, respectively, whereas GA (100 μM) decreased the activity by nearly 28% (Figure 4B), compared with water treatment (CK).

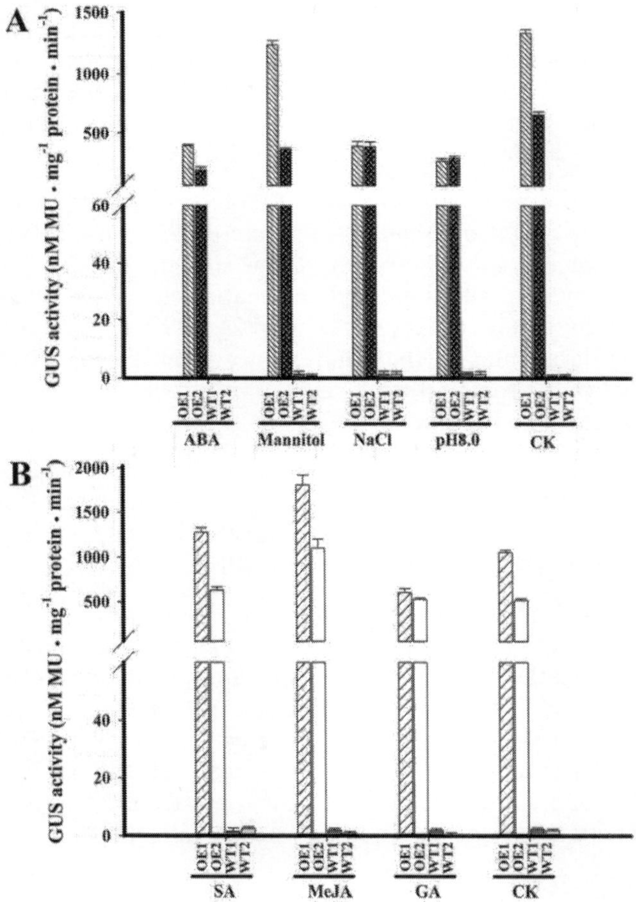

Figure 4. GUS activities of T_3 seedlings of Arabidopsis plant transformed with full-length BvcPPOP in the presence of abiotic stress signals. Data are shown as the mean value ± SD of three replicates of each BvcPPOP-GUS transgenic line. **(A)** Set 1 abiotic stress signals; **(B)** Set 2 abiotic stress signals.

2.5. 5′-Truncated BvcPPOPs Display Various Expression Patterns and/or Levels in Transgenic Arabidopsis

5′-Deletion of 70 bp at the far end of F1 (BvcPPOP) did not change the expression pattern of the GUS gene under its control at vegetative stages of transgenic Arabidopsis, but enhanced GUS staining and enzymatic activity in the root and petiole, in particular for 15-day-old seedlings and in petiole (F1 and F2 in Figure 2 and Figure 3). This enhancement resulted in a higher GUS activity in the petiole than that in root (Figure 3A). Although this enhancement extended to reproductive stages in the root and main veins of rosettes leaves, no stronger GUS-blue was noted in petiole (F1 and F2 in Figure 2). At reproductive stages, the F2-GUS plant showed light GUS blue in inflorescence shoot, but no GUS blue was visible in inflorescence leaf, flower, pod and seed.

Further 5′-deletion of 356 bp weakened GUS blue and enzymatic activity in the root of 5 to 10-day-old seedlings (F2 and F3 in Figure 2 and Figure 3), and no GUS blue was visible even at the root-tip of 15-day-old plants (Figure S1). However, GUS blue and enzymatic activity were more or less enhanced in the petiole of F3 (−1222 to +22)-GUS plants. At reproductive stages, the GUS staining in the inflorescence shoot remained unchanged, whereas that in the root and main veins of rosette leaves was significantly weakened (F2 and F3 in Figure 2).

Dramatic decline in GUS blue and enzymatic activity was noted in F4 (−803 to +22)-GUS plants from the vegetative stage to reproductive stage, but no change in GUS expression pattern was observed (F3 and F4 in Figure 2 and Figure 3), in comparison with the F3-GUS plant.

The weakest GUS blue was displayed in F5 (−225 to +22)-GUS seedlings. Although no GUS-blue was visible in 5-day-old plantlet, a faint GUS-blue was observed in the petiole and root of 10-day and older plants (F5 in Figure 2), and the GUS activity was detected (F5 in Figure 3). It is worth noting that at this vegetative stage, and even at such a low GUS activity, the GUS expression pattern remained unchanged, compared with F4. At reproductive stages, the GUS-blue weakened more and was visible only in the base of petiole and some sections of root (F5 in Figure 2).

3. DISCUSSION

A 1670 bp 5′-flanking region of a PPO gene was cloned from red Swiss chard, a betalain-producing plant (Figure 1), and as predicted, this region could direct expression of the foreign gene (GUS) in transgenic Arabidopsis, a non-betalain-producing plant (F1 in Figure 2), indicating its role as promoter. This promoter was named BvcPPOP.

Different from CaMV 35S promoter, the BvcPPOP (F1) drove expression of GUS in vegetative organs, but not or almost not in broad sense reproductive organs such as inflorescence shoot, inflorescence leaf, flower, pod and seed (F1 vs. 35S in Figure 2). Even among the vegetative organs, BvcPPOP-driven GUS expression was not equal. It expressed preferentially in root and petiole at the vegetative stage (F1 in Figure 2 and Figure 3). Furthermore, its expression was regulated developmentally. As plantlets growing and entering reproductive stage, the GUS activity in hypocotyl, petiole and main veins of rosette was almost absent or decreased significantly, although it remained unchanged in root (F1 in Figure 2).

In non-betalain plants, the PPO promoters were reported to be inducible by abiotic stress signals. For example, PINPPO1 promoter was induced by GA [13] and those of eggplant by MeJA and/or SA [11]. Our BvcPPOP cloned from the betalain plant, red Swiss chard, was also responsive to various abiotic signals tested. It was down-regulated by GA, ABA, NaCl and OH^- (pH 8.0) but up-regulated by SA and MeJA (Figure 4). This suggested that the native gene PPO under its control might be inducible by some abiotic stresses. What is worth to note is that these regulations were restricted to the expression strength, but not pattern.

In order to obtain the minimal promoter and elucidate the cis-acting elements controlling root- and/or petiole-preferential expression, we performed a set of 5′-deletions of BvcPPOP. The results showed that deletion of 70 bp at the 5′-extreme (from −1648 to −1579, noted as "F2") did not affect GUS expression pattern, but the expression strength in the petiole was increased and became higher than in root (F2 vs. F1 in Figure 2 and Figure 3B). This means that in the 70 bp fragment there may exist some kind of weak petiole-specific suppressor, which remains to be identified. Further deletion of 356 bp (from −1578 to −1223, noted as "F3") decreased more or less GUS expression strength in root, and in particular at the root tip and at reproductive stages, but not the expression pattern (F3 vs. F2 in Figure 2 and Figure 3B, Figure S1). In this region, there may be weak root-specific suppressor(s) that are up-regulated by the developmental stage. Apart from the suppressor(s), all cis-elements (Table S1) in this region seem functionally redundant. In contrast, deletion of the next 419 bp region (between −1222 bp and −804 bp) resulted in a dramatic decrement in GUS expression strength, although the expression pattern remained unchanged (F4 vs. F3 in Figure 2 and Figure 3). Accordingly, in this 419 bp region, there might exist some kind of "maintainer" or "enhancer" for gene expression. If the "maintainer" or "enhancer" was present, its function was much more pronounced in the region of −803 to −226, because deletion of this region left a very weak GUS expression and only in the petiole and root of 10-day and older plants at vegetative stages, and in the base of petiole and some sections of root at reproductive stages (F5 vs. F4 in Figure 2 and Figure 3). Interestingly, after a set of 5′-deletions, the rest of the region (−225 to + 22) of the BvcPPOP retained the capacity not only to drive GUS expression, but also to direct root- and petiole-preferential expression. Whether the root- and petiole-preferential expression is decided by cis-acting element(s) located upstream of the TATA-box or by 5′-UTR in this region remains to be elucidated, and which is under the progress.

It is well known that betalains seem inert to move from the tissues and/or organs where they are synthesized [32,36], and their synthesis is assumed to start by PPO-type tyrosinase-mediated tyrosine hydroxylation [24,26,28,29,37]. Thus, where the betalains accumulated and/or distributed must be the place where the promoter of gene(s) responsible for key enzyme(s) are involved in betalain biosynthesis and performs its function. In young red Swiss chard, the betalains are accumulated and/or distributed mainly in the root and petiole (Figure 5), and similarly, the strong GUS-blue appears essentially in the root and petiole in young plants of BvcPPOP-GUS transgenic Arabidopsis (Figure 5). This similarity or agreement implies that our cloned promoter BvcPPOP as well as its native gene BvcPPO might be involved in betalain biosynthesis.

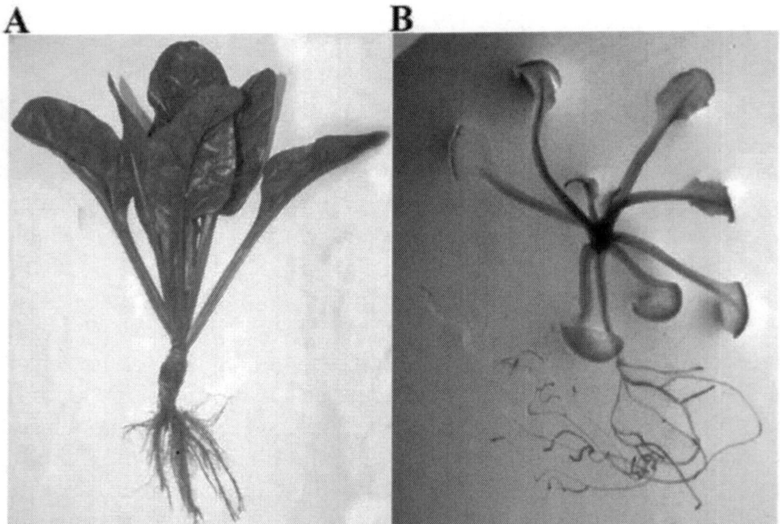

Figure 5. Comparison of betalain distribution and/or accumulation in young red Swiss chard with GUS staining pattern in young plants of BvcPPOP-GUS transgenic Arabidopsis. (**A**) Young red Swiss chard; (**B**) young plant of BvcPPOP-GUS transgenic Arabidopsis.

PPO's involvement in plant betalain biosynthesis is a long-standing hypothesis [24,26,28,29,37] and this hypothesis was challenged by the discovery of a new P450 (CYP76AD1) which could catalyze conversion of l-DOPA to cyclo-DOPA in sugar beet [36]. Hatlestad and co-workers (2012) even ruled out the involvement of conventional PPO and laccase based on relatively low abundance of the transcripts in sugar beet [36]. If this inference is correct in sugar beet, and applicable to red Swiss chard, our cloned promoter BvcPPOP may not be the promoter of such a conventional PPO, but that of secondary metabolism-specific PPO (Group 2 PPO) as described in Mayer, 2006 [38]; Thipyapong et al. 2007 [1] and Sullivan 2014 [2]. Thus, possible involvement of metabolism-specific PPO in betalain biosynthesis, especially in the first step and in the betaxanthin biosynthesis would be worth investigating.

In summary, a full-length PPO promoter (BvcPPOP) was cloned, for the first time, from a betalain-producing plant and this promoter could direct the GUS gene to express in the vegetative organs of an anthocyanin-producing plant, Arabidopsis, with a root- and petiole-preferential pattern. The promoter was regulated by developmental stage in its driving strength, but not its expression pattern. It was also regulated by various abiotic stressors tested. Its 5′-truncation versions varied remarkably in the driving strength, but not obviously in expression pattern, and even a short region from −225 to +22 retained root- and petiole-preferential expression pattern. This expression pattern is highly similar with that of betalain accumulation and/or distribution of red Swiss chard from which the promoter was cloned, implying possible involvement of the promoter's native gene in betalain biosynthesis, and particularly in the first step of the pathways and in betaxanthin biosynthesis. Successful cloning and functional elucidation of this PPO promoter from betalain-producing plants provides not only a useful tool for expressing gene(s) of agricultural interest in vegetative organs, but also an important clue to clarify the function of metabolism-specific PPO in betalain biosynthesis, which is still unclear.

4. EXPERIMENTAL SECTION

4.1. Plant Materials and Growth Conditions

Seeds of red and green Swiss chard (Beta vulgaris subspecies cicla) were kindly provided by Prof. Xi-Xiang Li of the Institute of Vegetables and Flowers, CAAS, China. Seeds were germinated in a plant growth incubator and then the seedlings were transplanted to pots in a greenhouse with natural light and temperature fluctuated from 20–25 °C during the day to 15–20 °C at night. Seeds of Arabidopsis thaliana (Columbia-0 ecotype) were surface-sterilized with 0.5% NaClO for 10 min and then rinsed five times with sterile water. The sterilized seeds were placed on MS medium [39]. Plates were transferred to a plant growth incubator for seed germination after vernalization at 4 °C for 3 days. The seedlings were planted in a culture room at 22 ± 1 °C with a photoperiod of 16-h light/8-h dark.

4.2. Isolation of the Promoter and Its Deletion Analysis

Genomic DNA was extracted from fresh leaves of red Swiss chard according to an improved CTAB (cetyltrimethylammonium bromide) method [40]. The promoter region was cloned by PCR with primer pair FW1/RW (Table 1), and the PCR products were inserted into pGEM-T vector (Promega, Madison, WI, USA) and sequenced. The resulted fragment, named BvcPPOP, was designed as full-length promoter (F1).

A set of 5′-deletions of F1 were generated by PCR using primer pairs FW2~FW5/RW (Table 1), and the truncated promoter fragments were also linked into the pGEM-T vector and sequenced as described above.

Table 1. Oligonucleotide primers used for PCR cloning and deletion of BvcPPOP promoter.

Primer Name	Primer Sequence (5′ to 3′) *
	Forward
FW1	AAGCTTGCAAATGGTCTATATCCGAC
FW2	AAGCTTGAGATAAACTCGATCAAGCTCG
FW3	AAGCTTGAGTAGTAGCAGTACAAGTTG
FW4	AAGCTTAGTATGTGTATGTCTTGAGTG
FW5	AAGCTTTACTAGTCGTAAAGTAGCAGTTCTC
	Reverse
RW	GGATCCTCTTTTATATATTTGGTGGA

* Underlined are *Hind* III and *Bam* HI sites at the forward and reverse primers, respectively.

4.3. Plant Expression Vector Construction and Agrobacterium-Mediated Transformation

The sequencing-verified promoter fragments were took out from their corresponding pGEM-T vector with Hind III and Bam HI and then individually cloned into a binary vector pBI121 (Clontech, Mountain View, CA, USA) to replace CaMV 35S promoter in front of β-glucuronidase (GUS) gene, which gave rise to pBvcPPOPn-GUS vectors (here n = 1 to 5).

All constructs including CaMV35S-GUS (original pBI121) were transferred into the Agrobacterium tumefaciens strain GV3101 by the freeze-thaw procedure [41]. Arabidopsis plants were transformed by using the floral dip method [42] and the T_1 transformants were selected on MS medium containing 50 mg/L of kanamycin (Kan). Kan-resistant seedlings were PCR verified and those with a strong and sharp PCR band were transplanted in pots and grown in a culture room. The T_3 generation of three independent lines for each construct was used for subsequent analysis, with wild type as negative controls.

4.4. Analysis of GUS Expression

Histochemical GUS staining and quantitative analysis for GUS activity in the transgenic plants were performed according to the method described by Jefferson et al. [43] and slightly modified by Han et al. [44]. Briefly, the Arabidopsis whole-plantlets of different developmental stages (5-day-old, 10-day-old, 15-day-old, after anthesis) were incubated at 37 °C for 12 to 16 h in GUS staining solution (50 mM phosphate buffer, pH 6.7, 1 mM EDTA pH 8.0, 0.2% (v/v) Triton-100, 1 mM K_3FeCN_6, 1 mM K_4FeCN_6, 0.5 mg/mL 5-bromo-4-chloro-3-indoxyl-d-glucuronic acid (X-gluc)). After staining, the samples were cleared in GUS destaining solution (75% ethanol, 25% acetic acid) for more than 1 h at room temperature, and then photographed using an Olympus SZX16-DP72 stereomicroscope. For stained plants of after-anthesis, they were photographed using a digital camera (Sony, DSLR-A350, Tokyo, Japan).

For quantitative GUS activity assay, the samples were prepared as previously described [45], and the enzymatic reaction was carried out according to Han et al. [44].

4.5. Abiotic Stress Treatment

To investigate the response of BvcPPOP to different abiotic stress signals, two independent F1-GUS transgenic lines (T_3 generation) were subjected to abiotic stressors. The stressors were divided in two sets: 1: NaCl, abscisic acid (ABA), mannitol, and OH^-; 2: gibberellin (GA), methyl jasmonate (MeJA) and salicylic acid (SA). For Set 1 stressor treatment, the disinfected seeds were vernalized at 4 °C for 3 days and then placed in a culture room with a photoperiod of 16-h light/8-h dark. Three-day-old plantlets of each line with a uniform growth were transplanted on the normal MS medium supplying with or without 120 mM NaCl, 2 μM ABA or, 200 mM mannitol or transplanted on higher-pH MS medium (pH 8.0). Seven days later, plants were sampled for protein extraction. For Set 2 stressor treatment, the roots of 10-day-old seedlings were immersed in water solution supplemented with or without 1 mM SA, 100 μM MeJA or 100 μM GA for 24 h [46]. GUS activity assay was performed as described above. The wild type plantlets were treated in parallel as negative controls. The mean values of two lines were used to estimate their regulation strength.

5. Conclusions

Red Swiss chard PPO promoter, BvcPPOP, with its 5'-truncted versions could direct expression of foreign gene(s) in vegetative organs with root- and petiole-preference, and this expression pattern is consistent with betalains accumulation/distribution in red Swiss chard. This promoter with its derivatives may be used in the gene-engineering for crop improvement and in the study of betalain biosynthesis.

Acknowledgments

We thank Xi-Xiang Li from Institute of Vegetables and Flowers, Chinese Academy of Agricultural Sciences (Beijing, China) for providing the seeds of red Swiss chard and green Swiss chard. We thank Ning Chen from our lab for the help on the plant expression vector construction. This work was supported by grants of Chinese High-Tech R & D Program "863" (Grant No. 2008AA10Z103) and National Natural Science Foundation of China (Grant No.31070268).

REFERENCES

1. Thipyapong, P.; Stout, M.J.; Attajarusit, J. Functional analysis of polyphenol oxidases by antisense/sense technology. Molecules **2007**, 12, 1569–1595.
2. Sullivan, M.L. Beyond brown: Polyphenol oxidases as enzymes of plant specialized metabolism. Front. Plant Sci. **2014**, 5, 783.
3. Li, L.; Steffens, J.C. Overexpression of polyphenol oxidase in transgenic tomato plants results in enhanced bacterial disease resistance. Planta **2002**, 215, 239–247.
4. Wang, J.H.; Constabel, C.P. Polyphenol oxidase overexpression in transgenic populus enhances resistance to herbivory by forest tent caterpillar (Malacosoma disstria). Planta **2004**, 220, 87–96.
5. Thipyapong, P.; Hunt, M.D.; Steffens, J.C. Antisense downregulation of polyphenol oxidase results in enhanced disease susceptibility. Planta **2004**, 220, 105–117.
6. Van Ranst, G.; Lee, M.R.F.; Fievez, V. Red clover polyphenol oxidase and lipid metabolism. Animal **2011**, 5, 512–521.
7. Wahler, D.; Gronover, C.S.; Richter, C.; Foucu, F.; Twyman, R.M.; Moerschbacher, B.M.; Fischer, R.; Muth, J.; Prufer, D. Polyphenoloxidase

silencing affects latex coagulation in taraxacum species. Plant Physiol. **2009**, 151, 334–346.

8. Taketa, S.; Matsuki, K.; Amano, S.; Saisho, D.; Himi, E.; Shitsukawa, N.; Yuo, T.; Noda, K.; Takeda, K. Duplicate polyphenol oxidase genes on barley chromosome 2H and their functional differentiation in the phenol reaction of spikes and grains. J. Exp. Bot. **2010**, 61, 3983–3993.
9. Thipyapong, P.; Steffens, J.C. Tomato polyphenol oxidase (differential response of the polyphenol oxidase F promoter to injuries and wound signals). Plant Physiol. **1997**, 115, 409–418.
10. Newman, S.M.; Tantasawat, P.; Steffens, J.C. Tomato polyphenol oxidase B is spatially and temporally regulated during development and in response to ethylene. Molecules **2011**, 16, 493–517.
11. Shetty, S.M.; Chandrashekar, A.; Venkatesh, Y.P. Promoter analyses and transcriptional profiling of eggplant polyphenol oxidase 1 gene (SmePPO1) reveal differential response to exogenous methyl jasmonate and salicylic acid. J. Plant Physiol. **2012**, 169, 718–730.
12. Richter, C.; Dirks, M.E.; Gronover, C.S.; Prufer, D.; Moerschbacher, B.M. Silencing and heterologous expression of PPO-2 indicate a specific function of a single polyphenol oxidase isoform in resistance of dandelion (Taraxacum officinale) against Pseudomonas syringae pv. tomato. Mol. Plant Microbe Interact. **2012**, 25, 200–210.
13. Zhou, Y.; O'Hare, T.J.; Jobin-Decor, M.; Underhill, S.J.; Wills, R.B.; Graham, M.W. Transcriptional regulation of a pineapple polyphenol oxidase gene and its relationship to blackheart. Plant Biotechnol. J. **2003**, 1, 463–478.
14. Naveed, I.; Mahmood, T. In silico characterization of rice polyphenol oxidase gene promoters. Arch. Sci. **2012**, 65, 297–304.
15. Quarta, A.; Mita, G.; Durante, M.; Arlorio, M.; de Paolis, A. Isolation of a polyphenol oxidase (PPO) cDNA from artichoke and expression analysis in wounded artichoke heads. Plant Physiol. Biochem. **2013**, 68, 52–60.
16. Li, D.; Deng, Z.; Liu, C.; Zhao, M.; Guo, H.; Xia, Z.; Liu, H. Molecular cloning, expression profiles, and characterization of a novel polyphenol oxidase (PPO) gene in Hevea brasiliensis. Biosci. Biotechnol. Biochem. **2014**, 78, 1648–1655.
17. Nakayama, T.; Yonekura-Sakakibara, K.; Sato, T.; Kikuchi, S.; Fukui, Y.; Fukuchi-Mizutani, M.; Ueda, T.; Nakao, M.; Tanaka, Y.; Kusumi, T.; et al.

Aureusidin synthase: A polyphenol oxidase homolog responsible for flower coloration. Science **2000**, 290, 1163–1166.

18. Kaintz, C.; Molitor, C.; Thill, J.; Kampatsikas, I.; Michael, C.; Halbwirth, H.; Rompel, A. Cloning and functional expression in E. coli of a polyphenol oxidase transcript from Coreopsis grandiflora involved in aurone formation. FEBS Lett. **2014**, 588, 3417–3426.
19. Cho, M.H.; Moinuddin, S.G.A.; Helms, G.L.; Hishiyama, S.; Eichinger, D.; Davin, L.B.; Lewis, N.G. (+)-Larreatricin hydroxylase, an enantio-specific polyphenol oxidase from the creosote bush (Larrea tridentata). Proc. Natl. Acad. Sci. USA **2003**, 100, 10641–10646.
20. Araji, S.; Grammer, T.A.; Gertzen, R.; Anderson, S.D.; Mikulic-Petkovsek, M.; Veberic, R.; Phu, M.L.; Solar, A.; Leslie, C.A.; Dandekar, A.M.; et al. Novel roles for the polyphenol oxidase enzyme in secondary metabolism and the regulation of cell death in walnut. Plant Physiol. **2014**, 164, 1191–1203.
21. Clement, J.; Mabry, T. Pigment evolution in the caryophyllales: A systematic overview. Bot. Acta **1996**, 109, 360–367.
22. Mueller, L.A.; Hinz, U.; Zrÿd, J.-P. Characterization of a tyrosinase from Amanita muscaria involved in betalain biosynthesis. Phytochemistry **1996**, 42, 1511–1515.
23. Steiner, U.; Schliemann, W.; Strack, D. Assay for tyrosine hydroxylation activity of tyrosinase from betalain-forming plants and cell cultures. Anal. Biochem. **1996**, 238, 72–75.
24. Strack, D.; Vogt, T.; Schliemann, W. Recent advances in betalain research. Phytochemistry **2003**, 62, 247–269.
25. Moreno, D.A.; García-Viguera, C.; Gil, J.I.; Gil-Izquierdo, A. Betalains in the era of global agri-food science, technology and nutritional health. Phytochem. Rev. **2008**, 7, 261–280.
26. Han, X.H.; Gao, Z.J.; Xiao, X.G. Enzymes and genes involved in the betalain biosynthesis in higher plants. Afr. J. Biotechnol. **2009**, 8, 6735–6744.
27. Brockington, S.F.; Walker, R.H.; Glover, B.J.; Soltis, P.S.; Soltis, D.E. Complex pigment evolution in the caryophyllales. New Phytol. **2011**, 190, 854–864.
28. Gandia-Herrero, F.; Garcia-Carmona, F. Biosynthesis of betalains: Yellow and violet plant pigments. Trends Plant Sci. **2013**, 18, 334–343. [Google Scholar] [CrossRef] [PubMed]

29. Khan, M.I.; Giridhar, P. Plant betalains: Chemistry and biochemistry. Phytochemistry **2015**, 117, 267–295.
30. Joy, R.W.T.; Sugiyama, M.; Fukuda, H.; Komamine, A. Cloning and characterization of polyphenol oxidase cDNAs of Phytolacca americana. Plant Physiol. **1995**, 107, 1083–1089.
31. Kaneda, H.; Matsui, T.; Tomiyasu, R.; Kuroda, Y.; Higashimoto, Y.; Oda, T.; Miyasaka, H.; Okuhata, H.; Tanaka, S.; Harada, K. Isolation of polyphenol oxidase genes from Portulaca oleracea and evaluation of their ability to metabolize endocrine-disrupting chemicals. Plant Biotechnol. **2012**, 29, 351–357.
32. Gao, Z.J.; Han, X.H.; Xiao, X.G. Purification and characterisation of polyphenol oxidase from red swiss chard (Beta vulgaris subspecies cicla) leaves. Food Chem. **2009**, 117, 342–348.
33. Xiao, X.G.; Gao, Z.J.; Han, X.H.; Liu, J.B.; Nie, X.L. Tyrosinase and its encoding gene in betalain-producing plants and their use thereof. China patent ZL201010263562.7, 2010.
34. Higo, K.; Ugawa, Y.; Iwamoto, M.; Korenaga, T. Plant cis-acting regulatory DNA elements (place) database: 1999. Nucleic Acids Res. **1999**, 27, 297–300.
35. Lescot, M.; Dehais, P.; Thijs, G.; Marchal, K.; Moreau, Y.; van de Peer, Y.; Rouze, P.; Rombauts, S. PlantCARE, a database of plant cis-acting regulatory elements and a portal to tools for in silico analysis of promoter sequences. Nucleic Acids Res. **2002**, 30, 325–327.
36. Hatlestad, G.J.; Sunnadeniya, R.M.; Akhavan, N.A.; Gonzalez, A.; Goldman, I.L.; McGrath, J.M.; Lloyd, A.M. The beet R locus encodes a new cytochrome p450 required for red betalain production. Nat. Genet. **2012**, 44, 816–820.
37. Tanaka, Y.; Sasaki, N.; Ohmiya, A. Biosynthesis of plant pigments: Anthocyanins, betalains and carotenoids. Plant J. **2008**, 54, 733–749. [Google Scholar] [CrossRef] [PubMed]
38. Mayer, A.M. Polyphenol oxidases in plants and fungi: Going places? A review. Phytochemistry **2006**, 67, 2318–2331.
39. Murashige, T.; Skoog, F. A revised medium for rapid growth and bio assays with tobacco tissue cultures. Physiol. Plant **1962**, 15, 473–497.
40. Paterson, A.H.; Brubaker, C.L.; Wendel, J.F. A rapid method for extraction of cotton (Gossypium spp.) genomic DNA suitable for RFLP or PCR analysis. Plant Mol. Biol. Rep. **1993**, 11, 122–127.

41. Chen, H.; Nelson, R.; Sherwood, J. Enhanced recovery of transformants of Agrobacterium tumefaciens after freeze-thaw transformation and drug selection. Biotechniques **1994**, 16, 664–668, 670.
42. Clough, S.J.; Bent, A.F. Floral dip: A simplified method for Agrobacterium-mediated transformation of Arabidopsis thaliana. Plant J. **1998**, 16, 735–743.
43. Jefferson, R.A.; Kavanagh, T.A.; Bevan, M.W. GUS fusions: β-Glucuronidase as a sensitive and versatile gene fusion marker in higher-plants. EMBO J. **1987**, 6, 3901–3907.
44. Han, L.; Han, Y.-N.; Xiao, X.-G. Truncated cotton subtilase promoter directs guard cell-specific expression of foreign genes in tobacco and arabidopsis. PLoS ONE **2013**, 8, e59802.
45. Geng, A.Q.; Zhao, Z.J.; Nie, X.L.; Xiao, X.G. Expression analysis of four flower-specific promoters of Brassica spp. in the heterogeneous host tobacco. Afr. J. Biotechnol. **2009**, 8, 5193–5200.
46. Chen, L.; Jiang, B.; Wu, C.; Sun, S.; Hou, W.; Han, T. GmPRP2 promoter drives root-preferential expression in transgenic arabidopsis and soybean hairy roots. BMC Plant Biol. **2014**, 14, 245.

8

CHAPTER

EFFECT OF EXTRACTION SOLVENT/TECHNIQUE ON THE ANTIOXIDANT ACTIVITY OF SELECTED MEDICINAL PLANT EXTRACTS

Bushra Sultana [1], Farooq Anwar [1,*] and Muhammad Ashraf [2]

[1]*Department of Chemistry and Biochemistry, University of Agriculture, Faisalabad-38040, Pakistan*

[2]*Department of Botany, University of Agriculture, Faisalabad-38040, Pakistan*

ABSTRACT

Theeffects of four extracting solvents [absolute ethanol, absolute methanol, aqueous ethanol (ethanol: water, 80:20 v/v) and aqueous methanol (methanol: water, 80:20 v/v)] and two extraction techniques (shaking and reflux) on the antioxidant activity of extracts of barks of Azadirachta indica, Acacia nilotica, Eugenia jambolana, Terminalia arjuna, leaves and roots of Moringa oleifera, fruit of Ficus religiosa, and leaves of Aloe barbadensis were investigated. The tested plant materials contained appreciable amounts of total phenolic contents (0.31-16.5 g GAE /100g DW), total flavonoid (2.63-8.66 g CE/100g DW); reducing power at 10 mg/mL extract concentration (1.36-2.91), DPPH· scavenging capacity (37.2-86.6%), and percent inhibition of linoleic acid (66.0-90.6%). Generally higher extract yields, phenolic contents and plant material antioxidant activity were obtained using aqueous organic solvents, as compared to the respective absolute organic solvents. Although higher extract yields were obtained by the refluxing extraction technique, in general higher amounts of total phenolic contents and better antioxidant activity were found in the extracts prepared using a shaker.

Keywords: medicinal plants; extraction effect; total phenolics; total flavonoids; antioxidant activity

INTRODUCTION

Plant-derived antioxidants, especially, the phenolics have gained considerable importance due to their potential health benefits. Epidemiological studies have shown that consumption of plant foods containing antioxidants is beneficial to health because it down-regulates many degenerative processes and can effectively lower the incidence of cancer and cardio-vascular diseases [1].

Recovery of antioxidant compounds from plant materials is typically accomplished through different extraction techniques taking into account their chemistry and uneven distribution in the plant matrix. For example, soluble phenolics are present in higher concentrations in the outer tissues (epidermal and sub-epidermal layers) of fruits and grains than in the inner tissues (mesocarp and pulp) [2]. Solvent extraction is most frequently used technique for isolation of plant antioxidant compounds. However, the extract yields and resulting antioxidant activities of the plant materials are strongly dependent on the nature of extracting solvent, due to the presence of different antioxidant compounds of varied chemical characteristics and polarities that may or may not be soluble in a particular solvent. Polar solvents are frequently employed for the recovery of polyphenols from a plant matrix. The most suitable of these solvents are (hot or cold) aqueous mixtures containing ethanol, methanol, acetone, and ethyl acetate [3]. Methanol and ethanol have been extensively used to extract antioxidant compounds from various plants and plant-based foods (fruits, vegetables etc.) such as plum, strawberry, pomegranate, broccoli, rosemary, sage, sumac, rice bran, wheat grain and bran, mango seed kernel, citrus peel, and many other fruit peels. Other studies have also demonstrated the efficacy of ethyl acetate to extract phenolic compounds from onion and citrus peel [3,4,5,6]. Bonoli et al. [7] reported that maximum phenolic compounds were obtained from barley flour with mixtures of ethanol and acetone. Similarly, aqueous methanol was found to be more effective in recovering highest amounts of phenolic compounds from rice bran [8], and Moringa oleifera leaves [9]. Anwar et al. [10] extracted antioxidant compounds from various plant materials including rice bran, wheat bran, oat groats and hull, coffee beans, citrus peel and guava leaves using aqueous 80% methanol (methanol: water, 80:20 v/v).

The medicinal plants selected for the present investigation, which included Moringa oleifera, Azadirachta indica, Terminalia arjuna, Acacia nilotica, Eugenia jambolana, Aloe barbadensis etc. have long been used in the folk medicine due to their potential health promoting and pharmacological attributes, which are mainly ascribed to the presence of antioxidant constituents such as phenolic acids and flavonoids [9,11,12,13]. It is important to establish appropriate means to evaluate and quantify effective antioxidant principles of medicinally or economically viable plant materials. The present study therefore was conducted with the main objective of investigating the most effective solvent/technique for extracting potent antioxidant compounds, especially phenolics from different parts of selected medicinal plants native to Pakistan.

RESULTS AND DISCUSSION

Effects of extracting solvent/technique on the extracts yields from different medicinal plant materials

Amounts (g/100g of dried plant material) of the antioxidant extract determined for different medicinal plant materials, using four different solvents (absolute methanol and aqueous methanol (methanol: water, 80:20 v/v); absolute ethanol and aqueous ethanol, (ethanol: water, 80:20 v/v) and two extracting techniques: shaker and reflux are shown in Table 1.

Our findings are in agreement with previous investigation of Chatha et al. [8], who reported that maximum extract yield (g/100g) from rice bran was obtained with aqueous methanol.

The differences in the extract yields from the tested plant materials in the present analysis might be ascribed to the different availability of extractable components, resulting from the varied chemical composition of plants [14]. The amount of the antioxidant components that can be extracted from a plant material is mainly affected by the vigor of the extraction procedure, which may probably vary from sample to sample. Amongst other contributing factors, efficiency of the extracting solvent to dissolve endogenous compounds might also be very important [9,15].

Table 1. Effects of extracting solvent/technique on the extract yield (g/100 g of DW) of medicinal plant materials.

Medicinal plant organs	Absolute methanol	Aqueous (80%) methanol	Absolute ethanol	Aqueous (80%) ethanol
	Extraction by shaker			
Moringa oleifera leaves	$9.61 \pm 0.39_d^c$	$17.9 \pm 0.18_c^a$	$8.94 \pm 0.27_d^c$	$12.6 \pm 0.51_d^b$
Moringa oleifera root	$3.24 \pm 0.14_e^b$	$6.65 \pm 0.19_e^a$	$2.23 \pm 0.12_e^b$	$3.63 \pm 0.26_e^b$
Eugenia jambolana bark	$19.2 \pm 0.38_b^a$	$14.1 \pm 0.56_d^b$	$2.81 \pm 0.39_e^c$	$13.5 \pm 0.40_d^b$
Acacia nilotica bark	$23.4 \pm 0.47_a^b$	$31.6 \pm 0.95_a^a$	$13.1 \pm 0.52_b^c$	$15.7 \pm 0.32_c^c$
Azadirachta indica bark	$10.7 \pm 0.22_d^c$	$13.8 \pm 0.55_d^c$	$37.2 \pm 0.74_a^a$	$25.0 \pm 0.53_b^b$
Terminalia arjuna bark	$22.5 \pm 0.67_a^b$	$23.3 \pm 0.45_b^b$	$34.5 \pm 0.44_a^a$	$37.2 \pm 0.46_a^a$
Ficus religiosa fruit	$18.9 \pm 0.76_b^b$	$26.4 \pm 0.52_b^a$	$16.9 \pm 0.67_b^b$	$19.7 \pm 0.39_c^b$
Aloe barbadensis leaves	$15.6 \pm 0.62_c^b$	$17.8 \pm 0.36_c^a$	$10.8 \pm 0.43_d^c$	$15.2 \pm 0.68_c^b$
	Extraction by reflux			
Moringa oleifera leaves	$16.6 \pm 0.33_d^b$	$21.1 \pm 0.84_c^a$	$12.2 \pm 0.37_c^c$	$17.2 \pm 0.35_c^b$
Moringa oleifera root	$5.12 \pm 0.21_e^{bc}$	$8.97 \pm 0.36_d^a$	$4.86 \pm 0.21_d^c$	$6.27 \pm 0.26_d^b$
Eugenia jambolana bark	$25.6 \pm 0.51_b^a$	$16.9 \pm 0.33_c^c$	$15.3 \pm 0.37_{bc}^c$	$19.5 \pm 0.46_c^b$
Acacia nilotica bark	$26.2 \pm 0.78_b^b$	$32.8 \pm 0.65_a^a$	$18.2 \pm 0.55_b^c$	$20.2 \pm 0.61_c^c$
Azadirachta indica bark	$14.2 \pm 0.29_d^c$	$17.8 \pm 0.53_c^c$	$42.4 \pm 0.64_a^a$	$31.9 \pm 0.63_b^b$
Terminalia arjuna bark	$28.6 \pm 0.46_a^b$	$24.9 \pm 0.49_b^b$	$40.7 \pm 0.86_a^a$	$46.6 \pm 083_a^a$
Ficus religiosa fruit	$21.3 \pm 0.64_c^c$	$29.2 \pm 0.88_a^a$	$19.5 \pm 0.58_b^c$	$22.8 \pm 0.91_c^c$
Aloe barbadensis leaves	$17.5 \pm 0.73_d^b$	$20.3 \pm 0.41_c^a$	$13.2 \pm 0.52_c^c$	$18.1 \pm 0.72_c^b$

Values (mean ± SD) are average of three samples of each medicinal plant material, analyzed individually in triplicate (n = 1x3 x 3), ($P < 0.05$); DW= dry weight; Superscript letters within the same row indicate significant ($P < 0.05$) differences of means within the extracting solvent; Subscript letters within the same column indicate significant ($P < 0.05$) differences of means within the plant materials.

For the effectiveness of extracting technique, the results showed that yields of the extract were better when extraction was done under reflux, regardless of the plant material and solvent used. This indicates that hot solvent systems under reflux state are more efficient for the recovery of antioxidant components, thus offering higher extract yields. This is in agreement with the findings of Shon et al. [16] who investigated that methanol and hot water are more efficient to extract antioxidant compounds from Phellinus baumii.

Effects of extracting solvent/technique on the total phenolic contents of different plant materials

Total phenolic contents (TPC) of different plant materials, using four solvent systems: absolute and aqueous methanol and absolute and aqueous ethanol and two extracting techniques (shaker and reflux) are presented in Table 2. Among the different medicinal plant materials, aqueous ethanolic extract of Acacia nilotica bark offered the highest TPC (16.5 g GAE/100g of DW), followed by aqueous ethanolic extract (aq. EE) of Terminalia arjuna bark (12.8 %), aq. ME of Moringa oleifera leaves (12.2%), aq. EE of Azadirachta indica bark (12.0%), aq. ME of Aloe barbadensis leaves

(10.3%), aq. EE of Eugenia jambolana bark (9.03%), aq. ME of Ficus religiosa fruit (5.34%), and aq. ME of Moringa oleifera roots (0.31%).

Table 2. Effects of extracting solvent/technique on the total phenolic contents (GAE g/100 g of DW) of medicinal plants materials.

Medicinal plant organs	**Absolute methanol**	**Aqueous (80%) methanol**	**Absolute ethanol**	**Aqueous (80%) ethanol**
	Extraction by shaker			
Moringa oleifera leaves	$10.3 \pm 0.41_{ab}^{ab}$	$12.2 \pm 0.28_{a}^{a}$	$9.72 \pm 0.21_{a}^{b}$	$11.6 \pm 0.21_{b}^{ab}$
Moringa oleifera root	$0.22 \pm 0.07_{d}^{b}$	$0.31 \pm 0.06_{e}^{a}$	$0.14 \pm 0.01_{e}^{c}$	$0.27 \pm 0.08_{f}^{ab}$
Eugenia jambolana bark	$10.1 \pm 0.39_{ab}^{a}$	$8.30 \pm 0.49_{b}^{b}$	$8.12 \pm 0.35_{b}^{b}$	$9.03 \pm 0.45_{c}^{ab}$
Acacia nilotica bark	$12.7 \pm 0.28_{a}^{b}$	$11.2 \pm 0.33_{ab}^{b}$	$11.2 \pm 0.31_{a}^{b}$	$16.5 \pm 0.66_{a}^{a}$
Azadirachta indica bark	$11.1 \pm 0.66_{ab}^{ab}$	$9.34 \pm 0.37_{ab}^{ab}$	$8.48 \pm 0.26_{b}^{b}$	$12.0 \pm 0.36_{b}^{a}$
Terminalia arjuna bark	$12.2 \pm 0.57_{a}^{a}$	$7.80 \pm 0.39_{c}^{b}$	$10.2 \pm 0.39_{a}^{ab}$	$12.8 \pm 0.26_{b}^{a}$
Ficus religiosa Fruit	$3.13 \pm 0.19_{c}^{ab}$	$5.34 \pm 0.36_{d}^{a}$	$2.67 \pm 0.16_{d}^{b}$	$4.11 \pm 0.18_{e}^{ab}$
Aloe barbadensis leaves	$8.25 \pm 0.28_{b}^{ab}$	$10.3 \pm 0.28_{ab}^{a}$	$6.53 \pm 0.38_{c}^{b}$	$7.93 \pm 0.31_{d}^{ab}$
	Extraction by reflux			
Moringa oleifera leaves	$9.63 \pm 0.28_{b}^{ab}$	$10.7 \pm 0.31_{a}^{a}$	$6.16 \pm 0.26_{c}^{b}$	$8.21 \pm 0.36_{c}^{ab}$
Moringa oleifera root	$0.17 \pm 0.02_{e}^{c}$	$0.27 \pm 0.04_{e}^{a}$	$0.12 \pm 0.03_{e}^{d}$	$0.23 \pm 0.06_{f}^{b}$
Eugenia jambolana bark	$8.91 \pm 0.39_{c}^{a}$	$8.14 \pm 0.33_{b}^{a}$	$7.94 \pm 0.31_{b}^{a}$	$8.64 \pm 0.27_{c}^{a}$
Acacia nilotica bark	$12.22 \pm 0.21_{a}^{ab}$	$10.7 \pm 0.24_{a}^{b}$	$10.8 \pm 0.28_{a}^{b}$	$14.6 \pm 0.29_{a}^{a}$
Azadirachta indica bark	$9.72 \pm 0.33_{b}^{a}$	$7.91 \pm 0.39_{b}^{b}$	$7.23 \pm 0.23_{b}^{b}$	$10.8 \pm 0.38_{b}^{a}$
Terminalia arjuna bark	$11.63 \pm 0.29_{ab}^{a}$	$6.25 \pm 0.30_{c}^{b}$	$9.67 \pm 0.38_{a}^{ab}$	$11.9 \pm 0.46_{b}^{a}$
Ficus religiosa Fruit	$2.12 \pm 0.09_{d}^{b}$	$4.93 \pm 0.28_{d}^{a}$	$2.26 \pm 0.10_{d}^{b}$	$4.13 \pm 0.21_{e}^{a}$
Aloe barbadensis leaves	$7.29 \pm 0.27_{c}^{b}$	$9.24 \pm 0.26_{ab}^{a}$	$6.44 \pm 0.29_{c}^{b}$	$6.94 \pm 0.27_{d}^{f}$

Values (mean ± SD) are average of three samples of each medicinal plant material, analyzed individually in triplicate (n = 1x3 x 3), ($P < 0.05$); DW= dry weight; Superscript letters within the same row indicate significant ($P < 0.05$) differences of means within the extracting solvent; Subscript letters within the same column indicate significant ($P < 0.05$) differences of means within the plant materials.

Results of the present study showed that among all the solvent extracts; the aqueous methanol and aqueous ethanol extracts had the highest TPC. This may be due to the fact that phenolics are often extracted in higher amounts in more polar solvents such as aqueous methanol/ethanol as compared with absolute methanol/ethanol [9,10,15].

The determined amounts of total phenolics (TP) from the tree barks investigated in the present study were lower than that reported for Acacia confusa bark [17]. Except for Eugenia jambolana, the barks of the other three plants offered greater amount of total phenolics than those of pine bark (11.4 g GCE/100g DW) [18]. TPC of Moringa oleifera leaves investigated in the present analysis are in agreement with previous reports [19]. The amount of TP of Moringa oleifera roots were found to be lower than those of Chinese herbal roots of kudzu vine (1.37 g GAE/100g) and dahurian (1.2 g GAE/100g) [20]. The levels of TP determined in the present analysis of Ficus religiosa fruit were found to be lower than those reported in Ficus microcarpa fruit (17.9 g GAE/100g) [21].

In contrast to the trends noted for extraction yields, the TPC of all medicinal plant materials extracted using the reflux technique decreased, regardless of the nature of the extracting solvent used. The decrease in the amounts of TP of these plant material extracts, prepared under reflux might have been due to the thermal decomposition of some phenolic antioxidants at the higher temperatures used for reflux extraction.

It has been reported that thermal processing conditions might result in the loss of natural antioxidants because heat may accelerate their oxidation and other degenerative reactions. Thus, heating temperature is of much consideration during processing. An accelerated shelf-life test at 80 °C for 4 days resulted in 20-40% decrease of the antioxidant activity of the apple juice [22]. Cheng et al. [23] reported that antioxidant activity of wheat bran decreased up to 61% by heating at 100 °C for 9 days. On the other hand, Dutra et al. [24] reported that among different extraction techniques (reflux, maceration, ultrasound, heating plate), extraction made under reflux using ethanol/water (70:30, v/v) offered the highest polyphenol levels in Vogel seeds. This might be attributed to an effective extraction under reflux conditions leading to higher release of some bound phenolics [2].

Effects of extracting solvent/technique on the total flavonoids of different plant materials

Total flavonoid contents (TFC) of various plant materials, extracted with four different solvent systems, using shaker and reflux extracting techniques, are given in Table 3. TFC were determined as catechin equivalents (CE). Among medicinal plant materials, aq. ME of Moringa oleifera leaves offered the highest TFC (8.66 g CE/100 g of DW) followed by aq. EE of Acacia nilotica bark (4.93), aq. ME of Aloe barbadensis leaves (2.28), aq. ME of Ficus religiosa fruit (3.77), aq. EE of Terminalia arjuna bark (3.49), aq. EE of Azadirachta indica bark (3.14), aq. ME of Moringa oleifera root(2.94), and ab. ME of Eugenia jambolana bark (2.63). Amount of TF in all the medicinal plant extracts generally decreased when reflux technique employed for their preparation. However, TFC of Aloe barbadensis leaves increased from 4.28 to 4.66 g CE/100 g of DW, when extracted with aqueous methanol using the reflux technique. Ficus religiosa fruits also contained higher TFC using the reflux technique with absolute and aqueous ethanol. TFC (1.47-3.77g/100g) of Ficus religiosa fruit in our analysis were found to be higher than that reported for Ficus microcarpa fruit (0.6 g/100 g dry weight) [21]. On the other hand, TFC in Terminalia arjuna bark (1.52-3.49 g/100g) determined in our work were lower than those (5.70 g/100 g dry weight) investigated by Dwivedi [12].

Table 3. Effects of extracting solvent/technique on the tctal flavonoid contents (CE g/100 g of DW) of medicinal plants materials.

Medicinal plant organs	Absolute methanol	Aqueous (80%) methanol	Absolute ethanol	Aqueous (80%) ethanol
	Extraction by shaker			
Moringa oleifera leaves	$6.06 \pm 0.12_a^b$	$8.66 \pm 0.21_a^a$	$5.33 \pm 0.13_a^b$	$6.21 \pm 0.11_a^b$
Moringa oleifera root	$1.68 \pm 0.06_d^b$	$2.94 \pm 0.08_c^a$	$1.22 \pm 0.04_e^b$	$1.59 \pm 0.03_d^b$
Eugenia jambolana bark	$2.63 \pm 0.04_c^a$	$1.72 \pm 0.07_d^{ab}$	$1.68 \pm 0.06_d^b$	$2.10 \pm 0.06_{cd}^{ab}$
Acacia nilotica bark	$4.86 \pm 0.09_b^a$	$3.21 \pm 0.12_{bc}^b$	$3.15 \pm 0.14_b^b$	$4.93 \pm 0.15_b^a$
Azadirachta indica bark	$2.93 \pm 0.10_c^a$	$3.31 \pm 0.16_{bc}^a$	$2.68 \pm 0.12_c^a$	$3.14 \pm 0.09_c^a$
Terminalia arjuna bark	$3.01 \pm 0.13_c^{ab}$	$2.13 \pm 0.13_c^b$	$2.64 \pm 0.09_c^{ab}$	$3.49 \pm 0.11_c^a$
Ficus religiosa fruit	$2.16 \pm 0.08_c^b$	$3.77 \pm 0.10_{bc}^a$	$1.28 \pm 0.04_e^c$	$2.03 \pm 0.06_{cd}^b$
Aloe barbadensis leaves	$2.91 \pm 0.16_c^b$	$4.28 \pm 0.17_b^a$	$1.68 \pm 0.02_d^c$	$2.96 \pm 0.04_c^b$
	Extraction by reflux			
Moringa oleifera leaves	$5.90 \pm 0.16_a^b$	$7.29 \pm 0.18_a^a$	$4.19 \pm 0.09_a^c$	$5.31 \pm 0.19_a^b$
Moringa oleifera root	$1.02 \pm 0.03_e^b$	$2.86 \pm 0.13_c^a$	$0.89 \pm 0.07_e^b$	$1.21 \pm 0.07_d^b$
Eugenia jambolana bark	$1.99 \pm 0.06_d^a$	$0.83 \pm 0.02_e^{bc}$	$1.06 \pm 0.04_{de}^c$	$1.55 \pm 0.05_d^b$
Acacia nilotica bark	$3.92 \pm 0.12_b^a$	$2.52 \pm 0.06_c^b$	$3.00 \pm 0.13_b^b$	$4.19 \pm 0.11_b^a$
Azadirachta indica bark	$2.16 \pm 0.08_d^b$	$2.90 \pm 0.04_c^a$	$1.66 \pm 0.06_d^c$	$2.99 \pm 0.13_c^a$
Terminalia arjuna bark	$1.78 \pm 0.06_d^{ab}$	$1.52 \pm 0.05_d^b$	$2.11 \pm 0.11_c^{ab}$	$2.63 \pm 0.08_c^a$
Ficus religiosa fruit	$1.97 \pm 0.05_d^c$	$3.56 \pm 0.11_{bc}^a$	$1.47 \pm 0.05_d^c$	$2.86 \pm 0.12_c^b$
Aloe barbadensis leaves	$2.90 \pm 0.07_c^b$	$4.66 \pm 0.09_b^a$	$1.39 \pm 0.07_d^c$	$2.55 \pm 0.09_c^b$

Values (mean ± SD) are average of three samples of each medicinal plant material, analyzed individually in triplicate (n = 1x3 x 3), ($P < 0.05$); DW= dry weight; Superscript letters within the same row indicate significant ($P < 0.05$) differences of means within the extracting solvent; Subscript letters within the same column indicate significant ($P < 0.05$) differences of means within the plant materials.

Effects of extracting solvent/technique on the reducing power of different plant materials

The results showing the effects of extracting solvent/techniques on the reducing potential of extracts of different plant materials at concentration of 10 mg/mL, are shown in Table 4. The reducing power of the medicinal plant extracts increased in a concentration dependent manner (data not shown). The values of absorbance for the tested extract solutions at concentration of 10 mg/mL determined in this assay, ranged from 0.09 to 2.88 and followed the order of effectiveness as: aq. ME of Moringa oleifera leaves (2.88)> aq. ME of Aloe barbadensis leaves (2.81) > aq. EE of Acacia nilotica bark (1.87)> aq. EE of Azadirachta indica bark (1.71) > aq. ME of Terminalia arjuna bark (1.66) > aq. EE of Eugenia jambolana bark (1.60) > aq. ME of Ficus religisa fruit (1.36) > aq. ME of Moringa oleifera roots (0.14).

In general, the aqueous organic solvent extracts of the tested plant materials, exhibiting greater TPC, also depicted good reducing power in the present analysis. The reducing potential of antioxidant components is very much associated with their TPC. The plant extracts with higher levels of total phenolics also exhibit greater reducing power [23,9,15].

As far as the effects of extraction techniques on the antioxidant activity is concerned, apart from the leaves of Aloe vera and fruit of Ficus religiosa, reducing powers of all the medicinal plant materials extracts were adversely affected by reflux extracting technique, regardless of the solvent used. However, each material tested retained the same efficacy order as displayed in the case of shaker extraction.

Table 4. Effects of extracting solvent/technique on reducing power (expressed as absorbance values at 700 nm) of different medicinal plant materials.

Medicinal plant organs	Absolute methanol	Aqueous (80%) methanol	Absolute ethanol	Aqueous (80%) ethanol
	Extraction by shaker			
Moringa oleifera leaves	$2.45 \pm 0.05_{a}^{b}$	$2.88 \pm 0.03_{a}^{a}$	$1.53 \pm 0.04_{a}^{c}$	$2.50 \pm 0.06_{a}^{b}$
Moringa oleifera root	$0.09 \pm 0.01_{e}^{b}$	$0.14 \pm 0.02_{c}^{a}$	$0.09 \pm 0.01_{c}^{b}$	$0.12 \pm 0.02_{d}^{a}$
Eugenia jambolana bark	$1.06 \pm 0.04_{d}^{c}$	$1.48 \pm 0.04_{b}^{b}$	$0.98 \pm 0.02_{b}^{c}$	$1.60 \pm 0.03_{bc}^{a}$
Acacia nilotica bark	$1.68 \pm 0.05_{c}^{ab}$	$1.52 \pm 0.04_{b}^{ab}$	$1.45 \pm 0.06_{a}^{b}$	$1.87 \pm 0.05_{b}^{a}$
Azadirachta indica bark	$1.55 \pm 0.03_{c}^{b}$	$1.46 \pm 0.05_{b}^{b}$	$1.05 \pm 0.05_{b}^{c}$	$1.71 \pm 0.03_{b}^{a}$
Terminalia arjuna bark	$1.26 \pm 0.02_{cd}^{ab}$	$1.66 \pm 0.04_{b}^{a}$	$1.12 \pm 0.02_{b}^{b}$	$1.34 \pm 0.04_{bc}^{ab}$
Ficus religiosa Fruit	$1.06 \pm 0.04_{d}^{b}$	$1.36 \pm 0.07_{b}^{a}$	$0.92 \pm 0.06_{b}^{b}$	$0.99 \pm 0.07_{cd}^{b}$
Aloe barbadensis leaves	$2.01 \pm 0.03_{b}^{b}$	$2.81 \pm 0.05_{a}^{a}$	$1.56 \pm 0.04_{a}^{c}$	$2.16 \pm 0.04_{a}^{b}$
	Extraction by reflux			
Moringa oleifera leaves	$1.25 \pm 0.05_{b}^{b}$	$1.78 \pm 0.03_{b}^{a}$	$0.94 \pm 0.04_{bc}^{c}$	$0.95 \pm 0.04_{c}^{c}$
Moringa oleifera root	$0.06 \pm 0.01_{d}^{b}$	$0.13 \pm 0.02_{d}^{a}$	$0.09 \pm 0.03_{d}^{b}$	$0.11 \pm 0.02_{d}^{a}$
Eugenia jambolana bark	$0.80 \pm 0.04_{c}^{b}$	$1.26 \pm 0.05_{cd}^{a}$	$0.61 \pm 0.05_{c}^{b}$	$1.39 \pm 0.07_{bc}^{a}$
Acacia nilotica bark	$1.25 \pm 0.06_{b}^{ab}$	$1.13 \pm 0.04_{d}^{b}$	$1.05 \pm 0.07_{bc}^{b}$	$1.62 \pm 0.05_{b}^{a}$
Azadirachta indica bark	$1.16 \pm 0.03_{b}^{b}$	$1.10 \pm 0.07_{d}^{b}$	$0.79 \pm 0.02_{c}^{b}$	$1.56 \pm 0.06_{b}^{a}$
Terminalia arjuna bark	$1.11 \pm 0.05_{b}^{b}$	$1.46 \pm 0.02_{c}^{a}$	$0.62 \pm 0.03_{c}^{c}$	$0.99 \pm 0.03_{c}^{b}$
Ficus religiosa Fruit	$1.13 \pm 0.02_{b}^{b}$	$1.22 \pm 0.05_{cd}^{a}$	$1.26 \pm 0.06_{b}^{a}$	$1.32 \pm 0.02_{bc}^{a}$
Aloe barbadensis leaves	$2.18 \pm 0.04_{a}^{b}$	$2.96 \pm 0.08_{a}^{a}$	$1.72 \pm 0.04_{a}^{c}$	$1.88 \pm 0.04_{a}^{c}$

Values (mean ± SD) are average of three samples of each medicinal plant material, analyzed individually in triplicate (n = 1x3 x 3), ($P < 0.05$); DW= dry weight; Superscript letters within the same row indicate significant ($P< 0.05$) differences of means within the extracting solvent; Subscript letters within the same column indicate significant ($P< 0.05$) differences of means within the plant materials.

Effects of extracting solvent/technique on the DPPH. Scavenging activity (% DPPH. remaining) of different plant materials

DPPH· scavenging activity of different plant materials as affected by extracting methods is depicted in Table 5. Absorbance in this assay was recorded at 0.5 to 10 min time intervals from initiation of the reaction. Observed scavenging activity was similar at the beginning of the reaction and changed with increase in the reaction time until it stabilized by the 10th min. Significant ($p< 0.05$) differences of DPPH· scavenging capacities among extracts were observed at 5th minute of the reaction. The DPPH· scavenging ability of the sample extracts was reported as the percent of DPPH·

scavenged (% DPPH· scavenging). As expected, a higher percent of DPPH· scavenging is correlated to a higher antioxidant activity [15,20].

Table 5. Effects of extracting solvent/technique the DPPH· scavenging activity (%) of different medicinal plant materials.

Medicinal plant organs	Absolute methanol	Aqueous (80%) methanol	Absolute ethanol	Aqueous (80%) ethanol
	Extraction by shaker			
Moringa oleifera leaves	$82.8 \pm 1.7_{a}^{a}$	$86.3 \pm 1.8_{a}^{a}$	$70.9 \pm 2.2_{b}^{b}$	$85.2 \pm 1.7_{a}^{a}$
Moringa oleifera root	$51.4 \pm 2.2_{cd}^{b}$	$62.9 \pm 1.6_{b}^{a}$	$56.6 \pm 1.2_{cd}^{ab}$	$61.8 \pm 2.3_{c}^{a}$
Eugenia jambolana bark	$48.8 \pm 1.4_{d}^{b}$	$52.7 \pm 1.3_{bc}^{a}$	$50.4 \pm 1.1_{d}^{ab}$	$53.9 \pm 1.9_{d}^{a}$
Acacia nilotica bark	$56.3 \pm 1.6_{c}^{b}$	$60.9 \pm 1.4_{b}^{b}$	$82.5 \pm 1.7_{a}^{a}$	$86.6 \pm 1.5_{a}^{a}$
Azadirachta indica bark	$53.2 \pm 1.5_{cd}^{b}$	$57.4 \pm 1.3_{b}^{ab}$	$57.3 \pm 1.4_{cd}^{ab}$	$60.8 \pm 1.8_{c}^{a}$
Terminalia arjuna bark	$37.7 \pm 1.3_{e}^{c}$	$48.5 \pm 1.5_{c}^{b}$	$62.9 \pm 1.5_{c}^{a}$	$67.4 \pm 1.4_{b}^{a}$
Ficus religiosa fruit	$57.2 \pm 1.7_{c}^{b}$	$63.4 \pm 1.1_{b}^{a}$	$55.3 \pm 1.3_{cd}^{b}$	$60.1 \pm 1.9_{c}^{a}$
Aloe barbadensis leaves	$73.7 \pm 1.3_{b}^{b}$	$80.1 \pm 2.3_{a}^{a}$	$67.2 \pm 1.9_{b}^{c}$	$70.7 \pm 1.2_{b}^{b}$
	Extraction by reflux			
Moringa oleifera leaves	$81.6 \pm 1.9_{a}^{a}$	$79.4 \pm 1.6_{a}^{b}$	$69.2 \pm 1.3_{b}^{c}$	$80.6 \pm 1.8_{a}^{a}$
Moringa oleifera root	$53.9 \pm 1.5_{c}^{b}$	$64.8 \pm 1.2_{b}^{a}$	$58.8 \pm 1.6_{c}^{ab}$	$62.7 \pm 1.9_{c}^{a}$
Eugenia jambolana bark	$48.1 \pm 1.7_{c}^{b}$	$51.4 \pm 1.4_{c}^{a}$	$49.2 \pm 1.2_{d}^{ab}$	$52.7 \pm 1.5_{d}^{a}$
Acacia nilotica bark	$55.8 \pm 1.3_{c}^{c}$	$60.7 \pm 1.6_{b}^{b}$	$81.9 \pm 1.7_{a}^{a}$	$81.1 \pm 1.8_{a}^{a}$
Azadirachta indica bark	$51.6 \pm 1.2_{c}^{b}$	$54.4 \pm 1.4_{c}^{ab}$	$57.7 \pm 1.3_{c}^{a}$	$58.4 \pm 1.7_{c}^{a}$
Terminalia arjuna bark	$37.2 \pm 1.1_{d}^{c}$	$47.3 \pm 1.3_{c}^{b}$	$61.6 \pm 1.3_{c}^{a}$	$65.9 \pm 1.4_{c}^{a}$
Ficus religiosa fruit	$55.9 \pm 1.4_{c}^{b}$	$62.9 \pm 1.7_{b}^{a}$	$56.2 \pm 1.4_{c}^{b}$	$63.8 \pm 1.6_{c}^{a}$
Aloe barbadensis leaves	$72.9 \pm 1.5_{b}^{ab}$	$77.6 \pm 1.9_{a}^{a}$	$68.0 \pm 1.3_{b}^{b}$	$71.9 \pm 1.2_{b}^{ab}$

Values (mean ± SD) are average of three samples of each medicinal plant material, analyzed individually in triplicate (n = 1x3 x 3), ($P < 0.05$); DW= dry weight; Superscript letters within the same row indicate significant ($P < 0.05$) differences of means within the extracting solvent; Subscript letters within the same column indicate significant ($P < 0.05$) differences of means within the plant materials.

The extracts of all the tested medicinal plant materials possessed free radical scavenging properties, but to varying degrees, ranging from 37.2 to 86.6% DPPH· scavenging. Using the shaker extraction technique, generally aq. EE and aq. ME showed better DPPH· scavenging activity. A maximum scavenging activity was offered by aq. EE of Acacia nilotica bark (86.6 %), followed by aq. ME of Moringa oleifera leaves (86.3%), aq. ME of Aloe barbadensis leaves (80.1%), aq. EE Terminalia arjuna bark (67.4%), aq. ME of Ficus religiosa fruit (63.4%), aq. ME of Moringa oleifera root (62.9%), aq. EE of Azadirachta indica bark (60.8%), and aq. EE of Eugenia jambolana bark (53.9%). The DPPH· scavenging activity seen in the different barks investigated in the present study was found to be lower than that of pine bark (95.1%) [17], while methanolic extract of Aloe barbadensis presented DPPH· scavenging activity (72.19%) comparable with that of earlier findings of Hu et al. [11].

The ethanolic extracts of roots of Moringa oleifera, leaves of Aloe barbadensis, and fruit of Ficus religiosa prepared by the reflux technique showed better scavenging activity as compared with those of prepared by the shaking technique. These results are in good agreement with the previous findings of Dutra et al. [23] who found that among different extraction techniques (reflux, maceration, ultrasound, heating plate) used, extraction made under reflux using ethanol/water (70:30, v/v) exhibited highest DPPH· scavenging activity. The rest of the medicinal plant materials extracts, prepared using shaker extracting method exhibited better scavenging activity than their corresponding extracts, obtained by reflux technique. It has well established that free radical scavenging activity of plant extracts is mainly due to phenolic compounds. This reduction in the radical scavenging activity of the extracts, obtained by the reflux technique might be ascribed to the thermal decomposition of phenolics [23].

Effects of extracting solvent/technique on the percent inhibition of linoleic acid peroxidation of different plant materials

Inhibition of linoleic acid oxidation determined for extracts of different plant materials as affected by extracting schemes are shown in Table 6.

Table 6. Effects of extracting solvent/technique on the percent inhibition of linoleic acid peroxidation of different medicinal plant materials.

Medicinal plant organs	Extraction by shaker			
	Absolute methanol	**Aqueous (80%) methanol**	**Absolute ethanol**	**Aqueous (80%) ethanol**
Moringa oleifera leaves	$79.9 \pm 2.4_{b}^{ab}$	$86.2 \pm 2.6_{a}^{a}$	$73.3 \pm 1.5_{b}^{b}$	$82.9 \pm 1.6_{b}^{a}$
Moringa oleifera root	$47.6 \pm 1.9_{e}^{b}$	$66.7 \pm 2.4_{d}^{a}$	$45.3 \pm 1.8_{e}^{b}$	$65.2 \pm 1.3_{c}^{a}$
Eugenia jambolana bark	$85.2 \pm 2.6_{a}^{ab}$	$80.4 \pm 1.6_{b}^{b}$	$90.6 \pm 2.7_{a}^{a}$	$90.2 \pm 1.8_{a}^{a}$
Acacia nilotica bark	$85.2 \pm 1.8_{a}^{a}$	$78.2 \pm 2.3_{b}^{b}$	$86.2 \pm 1.6_{a}^{a}$	$69.2 \pm 2.7_{c}^{c}$
Azadirachta indica bark	$65.0 \pm 2.6_{c}^{b}$	$71.4 \pm 2.8_{c}^{a}$	$47.8 \pm 1.8_{e}^{d}$	$55.1 \pm 1.6_{d}^{c}$
Terminalia arjuna bark	$31.1 \pm 1.2_{f}^{c}$	$44.4 \pm 1.7_{e}^{c}$	$61.8 \pm 2.4_{c}^{b}$	$66.0 \pm 2.6_{c}^{a}$
Ficus religiosa fruit	$59.2 \pm 1.7_{d}^{b}$	$67.4 \pm 2.1_{d}^{a}$	$54.9 \pm 2.1d_{d}^{c}$	$60.8 \pm 2.4_{d}^{b}$
Aloe barbadensis leaves	$66.2 \pm 1.3_{c}^{a}$	$68.3 \pm 1.4_{d}^{a}$	$63.7 \pm 1.9_{c}^{b}$	$65.9 \pm 1.9_{c}^{ab}$
	Extraction by reflux			
Moringa oleifera leaves	$68.2 \pm 2.0_{b}^{b}$	$82.6 \pm 1.6_{a}^{a}$	$68.7 \pm 2.0_{b}^{b}$	$80.5 \pm 1.7_{b}^{a}$
Moringa oleifera root	$40.8 \pm 1.6_{d}^{b}$	$64.8 \pm 2.5_{c}^{a}$	$43.2 \pm 1.7_{d}^{b}$	$63.6 \pm 1.9_{cd}^{a}$
Eugenia jambolana bark	$83.3 \pm 2.4_{a}^{a}$	$78.1 \pm 2.3_{a}^{b}$	$88.9 \pm 2.6_{a}^{a}$	$87.1 \pm 1.8_{a}^{a}$
Acacia nilotica bark	$84.1 \pm 1.5_{a}^{a}$	$72.1 \pm 1.4_{b}^{b}$	$85.3 \pm 2.5_{b}^{a}$	$66.3 \pm 1.6_{c}^{c}$
Azadirachta indica bark	$63.2 \pm 1.2_{b}^{b}$	$69.1 \pm 2.0_{bc}^{a}$	$44.3 \pm 1.7_{d}^{d}$	$50.8 \pm 2.0_{e}^{c}$
Terminalia arjuna bark	$59.4 \pm 1.7_{c}^{a}$	$36.2 \pm 1.4_{d}^{b}$	$61.4 \pm 1.2_{b}^{a}$	$63.2 \pm 2.5_{cd}^{a}$
Ficus religiosa fruit	$59.6 \pm 2.3_{c}^{b}$	$67.1 \pm 2.0_{bc}^{a}$	$55.2 \pm 1.6_{c}^{c}$	$61.9 \pm 2.4_{d}^{b}$
Aloe barbadensis leaves	$64.3 \pm 1.9_{b}^{b}$	$67.9 \pm 1.3_{bc}^{a}$	$66.2 \pm 1.3_{b}^{a}$	$67.3 \pm 2.1_{c}^{a}$

Values (mean ± SD) are average of three samples of each medicinal plant material, analyzed individually in triplicate (n = 1x3 x 3), ($P < 0.05$); DW= dry weight; Superscript letters within the same row indicate significant ($P < 0.05$) differences of means within the extracting solvent; Subscript letters within the same column indicate significant ($P < 0.05$) differences of means within the plant materials.

The present data were also compared with that of the synthetic antioxidant BHT (reference compound), which exhibited inhibition of linoleic acid oxidation at a level of 81.3 %. Among medicinal plant materials, maximum inhibition was noted by ab. EE of Eugenia jambolana bark (90.5%), followed byab. EE of Acacia nilotica bark= aq. ME of Moringa oleifera leaves (86.2%), aq. ME Azadirachta indica bark (71.4%), aq. ME of Aloe barbadensis leaves (68.3%), aq. ME of Ficus religiosa fruits (67.4%), aq. ME of Moringa oleifera roots (66.7%), and aq. EE of Terminalia arjuna bark (66.0%). The present level of percent inhibition exhibited by Moringa oleifera leaf extract (86.2%) was slightly lower than the values (89.7-92.0%) reported by Siddhuraju and Becker [9]. The extents of percent inhibition of linoleic acid (31.1-90.6%) exhibited by tree barks in the present investigation were found to be lower than the values (95.1%) offered by pine bark [17]. The present data revealed that, regardless of the solvent used, the extracts of all medicinal plant materials, prepared using the shaker extracting technique, exhibited higher levels of inhibition of linoleic acid oxidation than those obtained by the reflux method.

CONCLUSIONS

The results of the present investigation revealed that aqueous solvent (80% methanol, 80% ethanol) extracts of plant materials, prepared by both the shaker and reflux extraction techniques, exhibited better antioxidant activities and higher phenolic contents. Moreover, higher antioxidant extracts yields from the tested plant materials were obtained using the reflux extraction technique than by shaker, regardless of the solvent system used. Contrarily to extraction yield, the total phenolic contents and antioxidant activities of the tested plant materials decreased when these were extracted using the reflux technique. The present data would certainly help to ascertain the potency of the tested medicinal plant materials as potential source of natural antioxidants to be used for nutraceutical and functional food applications. However, further research is needed to identify individual components forming antioxidative system and develop their applications for food and pharmaceutical industries.

EXPERIMENTAL

Plant material

The selection of the plant materials in the present study was based on their potential medicinal uses. Medicinal plant parts i.e. barks of Azadirachta indica (Neem), Acacia nilotica (Desi kiker), Eugenia jambolana (Jaman),

Terminalia arjuna (Arjun), leaves and roots of Moringa oleifera (Sohanjana), fruit of Ficus religiosa (Peepal), and leaves of Aloe barbadensis (Aloe vera) were collected from plants in the vicinity of the University of Agriculture, Faisalabad, Pakistan. The subject plant material specimens were further authenticated by a taxonomist, Dr. Mansoor Hameed, Department of Botany, University of Agriculture, Faisalabad, Pakistan.

Chemicals and reagents

Analytical grade, Merck, Sigma and Fluka brand chemicals and reagents were used for the entire experimentation. 1,1-Diphenyl-2-picrylhydrazyl radical (DPPH·, Sigma, 90.0 %), linoleic acid, food grade synthetic antioxidant butylated hydroxytoluene (BHT, 99.0 %), Folin-Ciocalteu reagent (2 N), and gallic acid were purchased from Sigma Chemicals Co (St, Louis, MO, USA). All other chemicals (analytical grade) i.e. sodium hydroxide, sodium nitrite, ferrous chloride, ammonium thiocyanate, aluminum chloride, potassium dihydrogen phosphate, dipotassium hydrogen phosphate, used in this study were purchased from Merck (Darmstadt, Germany), unless stated otherwise.

Extraction of phenolic antioxidants

The air-dried ground (80 mesh) plant material (20 g for each sample) was extracted with each of the solvents – absolute ethanol, absolute methanol, aqueous ethanol (ethanol: water, 80:20 v/v) and aqueous methanol (methanol: water, 80:20 v/v) (200 mL) – for 6 hours at room temperature in an orbital shaker (Gallenkamp, UK), or under reflux on a water bath in separate experiments. The extracts were separated from the residues by filtering through Whatman No. 1 filter paper. The residues were extracted twice with the same fresh solvent and extracts combined. The combined extracts were concentrated and freed of solvent under reduced pressure at 45 °C, using a rotary evaporator (EYELA, SB-651, Rikakikai Co. Ltd. Tokyo, Japan). The dried crude concentrated extracts were weighed to calculate the yield and stored in a refrigerator (- 4 °C), until used for analyses.

Evaluation of antioxidant activity of plant materials/extracts

Determination of total phenolic content (TPC): Amount of TP were assessed using the Folin-Ciocalteu reagent [25]. Briefly, the crude extract (50 mg) was mixed with Folin-Ciocalteu reagent (0.5 mL) and deionized water (7.5 mL). The mixture was kept at room temperature for 10 min, and then 20% sodium carbonate (w/v, 1.5 mL) was added. The mixture was heated in a water bath at 40 °C for 20 min and then cooled in an ice bath; absorbance was read at 755 nm using a spectrophotometer (U-2001, Hitachi Instruments Inc., Tokyo, Japan). Amounts of TP were calculated using gallic acid calibration curve within range of 10-100 mgL^{-1}(R^2 = 0.9986). The results were expressed as gallic acid equivalents (GAE) g/100g of dry plant matter. All samples were analyzed thrice and the results averaged. The results are reported on dry weight basis (DW).

Determination of total flavonoid contents (TFC): The TFC were measured following a previously reported spectrophotometric method [26]. Briefly, extracts of each plant material (1 mL containing 0.1 mg/mL) were diluted with water (4 mL) in a 10 mL volumetric flask. Initially, 5% $NaNO_2$ solution (0.3 mL) was added to each volumetric flask; at 5 min, 10% $AlCl_3$ (0.3 mL) was added; and at 6 min, 1.0 M NaOH (2 mL) was added. Water (2.4 mL) was then added to the reaction flask and mixed well. Absorbance of the reaction mixture was read at 510 nm. TFC were determined as catechin equivalents (g/100g of dry weight). Three readings were taken for each sample and the results averaged.

Determination of reducing power: The reducing power of the extracts was determined according to the procedure described earlier [27], with a slight modification. Concentrated extract (2.5-10.0 mg) was mixed with sodium phosphate buffer (5.0 mL, 0.2 M, pH 6.6) and potassium ferricyanide (5.0 mL, 1.0%); the mixture was incubated at 50 °C for 20 min. Then 10% trichloroacetic acid (5 mL) was added and the mixture centrifuged at 980 g for 10 min at 5 °C in a refrigerated centrifuge (CHM-17; Kokusan Denki, Tokyo, Japan). The upper layer of the solution (5.0 mL) was decanted and diluted with 5.0 mL of distilled water and ferric chloride (1.0 mL, 0.1%), and absorbance read at 700 nm using a spectrophotometer (U-2001, Hitachi Instruments Inc., Tokyo, Japan). All samples were analyzed thrice and the results averaged.

DPPH· scavenging assay: 1, 1′-diphenyl–2-picrylhydrazyl (DPPH) free radical scavenging activity of the extracts was assessed using the procedure reported earlier [28]. Briefly, to extract (1.0 mL) containing 25 µg/mL of dry matter in methanol, freshly prepared solution of DPPH (0.025 g/L, 5.0 mL) was added. Absorbance at 0, 0.5, 1, 2, 5 and 10 min was measured at 515 nm using a spectrophotometer. The scavenging amounts of DPPH radical (DPPH·) was calculated from a calibration curve. Absorbance read at the 5th min was used for comparison of radical scavenging activity of the extracts.

Determination of antioxidant activity in linoleic acid system: The antioxidant activity of the tested plant extracts was also determined by measuring the oxidation of linoleic acid [28]. Five mg of each plant extract were added separately to a solution of linoleic acid (0.13 mL), 99.8% ethanol (10 mL) and 0.2 M sodium phosphate buffer (pH 7, 10 mL). The mixture was made up to 25 mL with distilled water and incubated at 40 °C up to 360 h. Extent of oxidation was measured by peroxide value applying thiocyanate method as described by Yen et al. [27]. Briefly, ethanol (75% v/v, 10 mL), aqueous solution of ammonium thiocyanate (30% w/v, 0.2 mL), sample solution (0.2 mL) and ferrous chloride ($FeCl_2$) solution (20 mM in 3.5% HCl; v/v, 0.2 mL) were added sequentially. After 3 min of stirring, the absorption was measured at 500 nm using a spectrophotometer (U-2001, Hitachi Instruments Inc., Tokyo, Japan). A control contained all reagents with exception of extracts. Synthetic antioxidants butylated hydroxytoluene (BHT) was used as a positive control. Percent inhibition of linoleic acid oxidation was calculated with the following equation: 100 – [(increase in absorbance of sample at 360 h / increase in absorbance of control at 360 h) × 100], to express antioxidant activity.

Statistical analysis

Three samples of each plant material were assayed. Each sample was analyzed individually in triplicate and data are reported as mean (n = 3 x 3 x1) ± SD (n = 3 x 3 x 1). Data were analyzed using a 2- way analysis of variance (ANOVA) using Minitab 2000 Version 13.2 statistical software (Minitab Inc. Pennsylvania, USA). A probability value of $p \leq 0.05$ was considered to denote a statistically significance difference.

REFERENCES

1. Arabshahi-Delouee, S.; Urooj, A. Antioxidant properties of various solvent extracts of mulberry (Morus indica L.) leaves. Food Chem. **2007**, 102, 1233–1240.
2. Antolovich, M.; Prenzler, P.; Robards, K.; Ryan, D. Sample preparation in the determination of phenolic compounds in fruits. Analyst **2000**, 125, 989–1009.
3. Peschel, W.; Sanchez-Rabaneda, F.; Dn, W.; Plescher, A.; Gartzia, I.; Jimenez, D.; Lamuela-Raventos, R.; Buxaderas, S.; Condina, C. An industrial approach in the search of natural antioxidants from vegetable and fruit wastes. Food Chem. **2006**, 97, 137–150.
4. Abdille, M.H.; Singh, R.P.; Jayaprakasa, G.K.; Jens, B.S. Antioxidant activity of the extracts from Dillenia indica fruits. Food Chem. **2005**, 90, 891–896.
5. Rehman, Z.U. Citrus peel extract- A natural source of antioxidant. Food Chem. **2006**, 99, 450–454.
6. Li, Y.; Guo, C.; Yang, J.; Wei, J.; Xu, J.; Cheng, S. Evaluation of antioxidant properties of pomegranate peel extract in comparison with pomegranate pulp extract. Food Chem. **2006**, 96, 254–260.

7. Bonoli, M.; Verardo, V. ; Marconi, E. ; Caboni, M.F. Antioxidant phenols in barley (Hordeum vulgare L.) flour: comparative spectrophotometric study among extraction methods of free and bound phenolic acids. J. Agric. Food Chem. **2004**, 52, 5195–5200.
8. Chatha, S.A.S.; Anwar, F.; Manzoor, M.; Bajwa, J.R. Evaluation of the antioxidant activity of rice bran extracts using different antioxidant assays. Grasas Aceites Sevilla **2006**, 57, 328–335.
9. Siddhuraju, P.; Becker, K. Antioxidant properties of various extracts of total phenolic constituents from three different agroclimatic origins of drumstick tree (Moringa oleifera lam.) leaves. J. Agric. Food Chem. **2003**, 51, 2144–2155.
10. Anwar, F.; Jamil, A.; Iqbal, S.; Sheikh, M.A. Antioxidant activity of various plant extracts under ambient and accelerated storage of sunflower oil. Grasas Aceites Sevilla **2006**, 57, 189–197.
11. Hu, Y.; Xu, J. u, Q, Evaluation of Antioxidant Potential of Aloe vera (Aloe barbadensis Miller) Extracts. J. Agric. Food Chem. **2003**, 51, 7788–7791.
12. Dwievedi, S. Terminalia arjuna Wight & Arn.-A useful drug for cardiovascular disorders. J. Ethnopharm. **2007**, 114, 114–129.
13. Sultana, B.; Anwar, F. Flavonols (kaempferol, quercetin, myricetin) contents of selected fruits, vegetables and medicinal plants. Food Chem. **2008**, 108, 879–884.
14. Hsu, B.; Coupar, I.M.; Ng, K. Antioxidant activity of hot water extract from the fruit of the Doum palm, Hyphaene thebaica. Food Chem. **2006**, 98, 317–328.
15. Sultana, B.; Anwar, F.; Przybylski, R. Antioxidant activity of phenolic components present in barks of barks of Azadirachta indica, Terminalia arjuna, Acacia nilotica, and Eugenia jambolana Lam. trees. Food Chem. **2007**, 104, 1106–1114.
16. Shon, M.Y.; Choi, S.D. ; Kohng, G.G. ; Nam, S.H ; Sung, N.J. Antimutagenic, antioxidant and free radical scavenging activity of ethyl acetate extracts from white, yellow and red onion. Food Chem. Toxicol. **2004**, 42, 659–666.
17. Tung, Y.T.; Wu, J.H.; Kuo, Y.H.; Chang, S.T. Antioxidant activities of natural phenolic compounds from Acacia confusa bark. Bicres. Technol. **2007**, 98, 1120–1123.
18. Yu, L.; Zhao, M.; Wang, J.S.; Cui, C.; Yang, B.; Jiang, Y.; Zhao, Q. Antioxidant, immunomodulatory and anti-breast cancer activities of

phenolic extract from pine (Pinus massoniana Lamb) bark. Inno. Food Sci. Emer. Technol. **2008**, 9, 122–128.

19. Iqbal, S.; Bhanger, M.I. Effect of season and production location on the antioxidant activity of Moringa oleifera leaves grown in Pakistan. J. Food Comp. Anal. **2006**, 102, 544–551.
20. Liu, H.; Qiu, N.; Ding, H.; Yao, R. Polyphenols contents and antioxidant capacity of 68 Chinese herbals suitable for medicinal or food uses. Food Res. Inter. **2008**, 41, 363–370.
21. Ao, C.; Li, A.; Elzaawely, A.A.; Xuan, T.D.; Tawata, S. Evaluation of antioxidant and antimicrobial activities of Ficus microcarpa L. fil. extract. Food Contr. **2008**, 19, 940–948. [Google Scholar] [CrossRef]
22. Van der Sluis, A.A.; Dekker, M.; Boekel, M.A.J.S. Activity and concentration of polyphenolic antioxidants in apple juice. 3. stability during storage. J. Agric. Food Chem. **2005**, 53, 1073–1080.
23. Cheng, Z.; Su, L.; Moore, J.; Zhou, K.; Luther, M.; Yin, J.J.; Yu, L.L. Effect of postharvest treatment and heat stress on availability of wheat antioxidants. J. Agric. Food Chem. **2006**, 54, 5623–5629.
24. Dutra, R.C.; Leite, M.N.; Barbosa, N.R. Quantification of phenolic constituents and antioxidant activity of Pterodon emarginatus vogel seeds. Inter. J. Mol. Sci. **2008**, 9, 606–614.
25. Chaovanalikit, A.; Wrolstad, R.E. Total anthocyanins and total phenolics of fresh and processed cherries and their antioxidant properties. J. Food Sci. **2004**, 69, 67–72.
26. Dewanto, V.; Wu, X.; Adom, K.K.; Liu, R.H. Thermal processing enhances the nutritional value of tomatoes by increasing total antioxidant activity. J. Agric. Food Chem. **2002**, 50, 3010–3014.
27. Yen, G.C.; Duh, P.D.; Chuang, D.Y. Antioxidant activity of anthraquinones and anthrone. Food Chem. **2000**, 70, 307–315.
28. Iqbal, S.; Bhanger, M.I.; Anwar, A. Antioxidant properties and components of some commercially available varieties of rice bran in Pakistan. Food Chem. **2005**, 93, 265–272.

9

CHAPTER

MULTI-INSTRUMENTAL ANALYSIS OF TISSUES OF SUNFLOWER PLANTS TREATED WITH SILVER(I) IONS – PLANTS AS BIOINDICATORS OF ENVIRONMENTAL POLLUTION

Rene Kizek et al.

[1]*Department of Chemistry and Biochemistry, Mendel University of Agriculture and Forestry, Zemedelska 1, CZ-613 00 Brno, Czech Republic*

ABSTRACT

The aim of this work is to investigate sunflower plants response on stress induced by silver(I) ions. The sunflower plants were exposed to silver(I) ions (0, 0.1, 0.5, and 1 mM) for 96 h. Primarily we aimed our attention to observation of basic physiological parameters. We found that the treated plants embodied growth depression, coloured changes and lack root hairs. Using of autofluorescence of anatomical structures, such as lignified cell walls, it was possible to determine the changes of important shoot and root structures, mainly vascular bungles and development of secondary thickening. The differences in vascular bundles organisation, parenchymatic pith development in the root centre and the reduction of phloem part of vascular bundles were well observable. Moreover with increasing silver(I) ions concentration the vitality of rhizodermal cells declined; rhizodermal cells early necrosed and were replaced by the cells of exodermis. Further we employed laser induced breakdown spectroscopy for determination of spatial distribution of silver(I) ions in tissues of the treated plants. The Ag is accumulated mainly in near-root part of the sample. Moreover basic biochemical indicators of environmental stress were investigated. The total content of proteins expressively decreased with increasing silver(I) ions dose and the time of the treatment. As we compare the results obtained by protein analysis – the total protein contents in shoot as well as root parts – we can assume on the transport of the proteins from the roots to shoots. This phenomenon can be related with the cascade of processes connecting with photosynthesis. The second biochemical parameter, which we investigated, was urease activity. If we compared the activity in treated plants with control, we found out that presence of silver(I) ions markedly enhanced the activity of urease at all applied doses of this toxic metal. Finally we studied the effect of silver(I) ions on activity of urease in in vitro conditions.

Keywords: Silver; Heavy metals; Plant biosensor; Sensors; Biochemical marker

1. INTRODUCTION

1.1 Silver ions and their effects on organisms

Due to many anthropogenic activities environment have been polluting by number of organic as well as inorganic compounds [1]. Therefore the concentration of these undesirable and in most cases highly toxic substances have enhanced [2]. The mechanism of their effects can be very heterogeneous. Because of these reasons the new procedures and technologies not only to monitor of levels of contamination but also to remediate the polluted environment have been still developing and suggesting. The ions of heavy metals and their compounds are considered as one of the most toxic substances polluting all parts of environment. Photographical industry, following electrochemistry and medicine are the main sources of one of the most toxic heavy metal ions, silver(I) ions. In water silver(I) can be found in many forms both as inorganic and organic compounds. Due to competing equilibria and kinetics in water hydrated silver ions, Ag^+, may be also present in surface waters, which relates to the fact that Ag^+ has been shown to be highly toxic to aquatic life [3-5]. In the light of these facts the foundation called "The Silver Council", which was established in 1998, elaborated the basic characteristics of environmental norms for silver ions entering to environment from industry [6]. However the toxic effect of silver(I) ions on the organisms is still unclear.

There have been suggested several mechanisms of silver ions effects on vitally important functions and processes. One of the most discussed is the interactions between Ag^{+} and DNA, which have been studied by number of analytical methods (infrared and UV spectrometry, circular dichroism etc.) [7,8]. It was observed that silver ions bind rapidly on N7 guanine in dsDNA and, thus, interfere the replication processes [8].

It is generally known that prokaryotic as well as eukaryotic organisms intensively protect themselves against effects of heavy metals. The both groups of organisms have developed various strategies how they can detoxify xenobiotics. One of the common ones is synthesis of low molecular peptides and proteins rich in cysteine. In general plants synthesize peptides called phytochelatins [9,10] and animals low molecular proteins called metallothioneins [11].

1.2 Bio-indicators

Numerous of plant and animal species can be used as bio-indicators of heavy metals pollution of the environment [12-19]. Aquatic animals, most of all various species of fishes, are very suitable for these purposes [20-22]. Nevertheless the question is how to asses the quality of the environment of interest through using of such living bio-indicators [23]. To assess the quality of the environment the investigations of the changes in behaviour, morphology, habitation or changes of basic morphometric properties (body weight, colour, length, etc.) are often used. All these data are only of qualitative character and are hardly to obtain due to requirements on large population of the target specie and the time period of the experiment. On the other hand, experiments using plants can be appropriate way how to substitute animal tests because no harming of animals and low demands on equipment.

1.3 Silver ions analysis

A determination of silver(I) ions in waters is difficult because the formation of a number of silver complexes with inorganic as well as organic compounds that however depress the acute silver toxicity [6,24,25]. The determination of silver ions is usually carried out by atomic absorption spectrometry [26,27]. To enhance the sensitivity of an analysis the pre-concentration of the silver ions in a sample is need. These processes prolong the total time of the analysis as well as enhance the cost of such experiment [27-29]. The electrochemical methods are alternative analytic techniques that make the silver ions determination possible in nM concentrations, mainly using carbon electrodes [30-36]. The modification of the surface of the carbon electrodes represents a unique tool for detection of heavy metal ions, peptides, proteins, nucleic acids and others [37-46].

The aim of this work is to investigate sunflower plants response on stress induced by silver(I) ions. For this purpose we employed multi-instrumental apparatus to detect and investigate total protein content, urease activity, spatial distribution of the heavy metal ions, and physiological and anatomical changes in the treated plants.

2. MATERIAL AND METHODS

2.1 Chemicals and pH measurements

Urease EC 3.5.1.5 (Jack Beans, type III; 45 000 IU/g) was purchased from Sigma Aldrich (St. Louis, USA). Acetic acid was purchased from Fluka (USA). All other reagents used were purchased from Sigma Aldrich in ACS purity unless noted otherwise. Stock standard solutions were prepared by ACS water (Sigma-Aldrich, USA) and stored in the dark at temperature of -20 °C. Working standard solutions were prepared daily by dilution of the stock solutions. All solutions were filtered through a 0.45 μm Nylon filter discs (Millipore, Billerica, Mass., USA) prior to HPLC analysis. The pH value was measured using WTW inoLab Level 3 with terminal Level 3 (Weilheim, Germany), controlled by the personal computer program (MultiLab Pilot; Weilheim, Germany). The pH-electrode (SenTix-H, pH 0–14/3M KCl) was regularly calibrated by set of WTW buffers (Weilheim, Germany).

2.2 Plants, cultivation conditions and a sample preparation

Sunflower plants (Helianthus annuus L.) were used in our experiments. The sunflower seeds were germinated on wet filter paper in special vessels at 25 ± 2 °C in dark (box Chirana, Czech Republic). After 10 days, the sunflower seedlings were placed into vessels containing tap water and cultivated in Versatile Environmental Test Chamber (MLR-350 H, Sanyo, Japan) for eight days with 14 hours long daylight per day (maximal light intensity was about 100 $\mu E.m^{-2}s^{-1}$) at a temperature 22 °C and humidity 65 %. Further $AgNO_3$ was added to the cultivation solution at final concentrations of 0 (control sample), 0.1, 0.5 and 1 mM. The sunflower plants placed in the vessels that contained tap water with addition of silver(I) ions were treated for five days. Seven plants from each experimental group were harvested at certain time intervals during the experiment, and their roots were rinsed three times in distilled water and 0.5 M EDTA. Prior to their analysis each harvested plant was divided into leaves, stem and root.

2.3 Laser spectrometry

The LIBS experimental setup used is shown in Fig. 1. The second harmonic (532 nm) of the Nd:YAG laser system (Brilliant B, Quantel, France) was used to create the LIBS micro-plasma by focusing the laser beam with a 16 mm focal-length glass doublet (Sill Optics, Germany). The laser pulse width was ~ 5 ns and beam diameter 8 mm. The energy of the laser pulse (~10 mJ at the sample) was set and controlled by an energy meter (Field Master LM-P10, Coherent, USA).

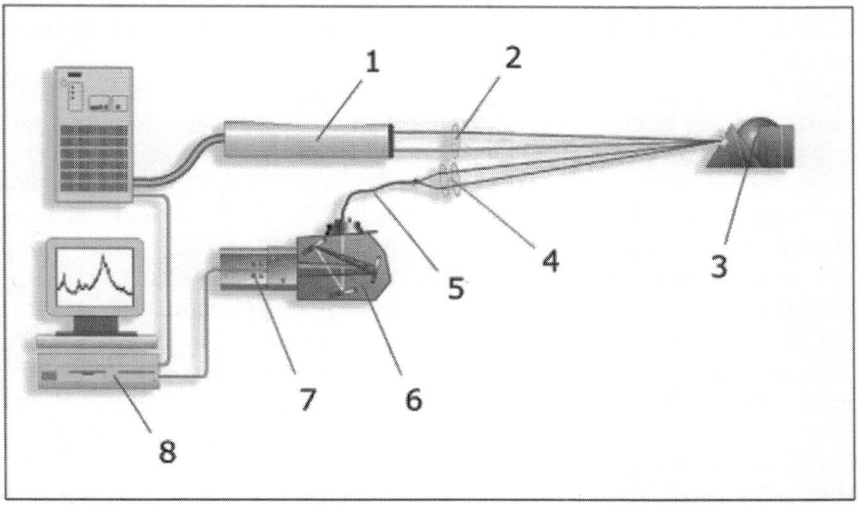

Figure 1. The LIBS experimental setup. 1 – Nd:YAG ablation laser, 2 – focusing optics, 3 – the analyzed sample, 4 – collecting optics, 5 – optical fiber, 6 – monochromator, 7 – ICCD camera, 8 – personal computer.

The sample was placed to the sample holder inside the ablation chamber (Tescan, Czech Republic) to the stage with precision movements (2 µm in x, y and z direction). The LIBS analysis was performed in air on atmospheric pressure. The ablation spot was targeted and controlled for each shot by a CCD camera placed outside of the chamber. For the temporally and spectrally resolved analysis the LIBS plasma radiation was collected with quartz objectives and transported by a 3 m fiber optic system onto the entrance slit of the 0.32 m monochromator (TRIAX 320, Jobin Yvon, France). In this study, the grating 2400 g/mm of the monochromator and 50 µm entrance slit were used. As a detector an ICCD camera (Horiba, Jobin Yvon, France) was employed. The camera was triggered by the Q-switch signal of the laser.

2.4 Automated spectrometric measurements

Spectrometric measurements were carried using an automated chemical analyser BS-200 (Mindray, China). Reagents and samples were placed on cooled sample holder (4 °C) and automatically pipetted directly into plastic cuvettes. Incubation proceeded at 37°C. Mixture was consequently stirred. The washing steps by distilled water (18 mΩ) were done in the midst of the pipetting. Apparatus was operated using software BS-200 (Mindray, China;.

2.4.1 Urease activity determination – An indophenol assay (Berthelot method)

Plant tissues samples (approximately 2 grams) were homogenized in mortar for five minutes. Then twenty millilitres of 30% ethanol was added and this solution was poured into a bottle (50 ml) and vortexed at 300 rpm, 8 °C for 30 minutes using vortex (GFL, Germany). The extract was centrifuged for 10 min at 5 000 g (Hettich, Geramy) ànd then the supernatant was collected. The supernatant (10 μl) was mixed with 448 μl of hypochlorite solution (12% NaOCl, 0.4 M Na_2HPO_4 and 0.37 M NaOH, adjusted to pH 12) and with 42 μl of phenol solution (sodium nitroprusside, 7% phenol). This mixture was stirred and incubated for 15 min at 37°C. After this incubation the differences of absorption at 630 and 670 nm were measured.

2.4.2 Protein determination – Biuret test

Weighed plant tissues (approximately 0.2 g) were transferred to a test-tube. Then, liquid nitrogen was added to the test-tube, and the samples were frozen to disrupt the cells. The frozen sample was transferred to mortar and spread for 1 min. Then exactly 1,000 μl of 0.2 M phosphate buffer (pH 7.2) was added to mortar, and the sample was spread for 5 min. The homogenate was transferred to a new test-tube. The frozen samples were homogenised by shaking on a Vortex–2 Genie for 5 min at 4 °C (Scientific Industries, USA) and sonicated using a Bandelin Sonopuls HD 2070 for 10 s at 7 W (Germany). The homogenate was centrifuged (14 000 × g) for 15 min at 4°C using a Universal 32 R centrifuge (Hettich-Zentrifugen, Germany). The supernatant was filtered through a 0.45 μm Nylon filter discs (Millipore, Billerica, Mass., USA) prior to analysis.

For determination of the total protein content the biuret solution (15 mM potassium sodium tartrate, 100 mM NaI, 15 mM KI and 5 mM $CuSO_4$). As a standard albumin (1 mg in 1 ml of phosphate buffer, pH 7) was used. The measurement was done as follows: 180 μl of the biuret solution was mixed with 45 μl of real or standard sample; after stirring and incubating (10 min. at 37°C) the absorbance at 546 nm was measured.

2.5 Anatomical analysis of plant samples

The observations were performed with fresh stem sections of plants. Lignified cell walls were visualized by ultraviolet (UV) excitation without additional fluorescent staining. At each experimental group of plants 25 – 30 sections were evaluated. All observations were performed on fluorescent microscope (Olympus AX 70) equipped with broad spectrum UV excitation.

2.6 Statistical analysis

Data were processed using MICROSOFT EXCEL® (USA) and analyzed by the QCExpert software (TriloBite, Statistical Software, Czech Republic) using analysis of variance (ANOVA). Results are expressed as mean ± S.D. unless noted otherwise. Statistical significance of the differences between weight and area of clusters was determined. Differences with $p < 0.05$ were considered significant.

3. RESULTS AND DISCUSSION

In the case of silver ions toxicity the attention is devoted to aquatic organisms [1,47-53], mainly to fishes due to extremely low lethal doses, which vary from units to ten of micrograms per litre [3,4,48-56]. On the other hand it is little known about the influence of silver ions on plants. It is well known that a presence of a toxic metal in a plant cell results in synthesis of stress plant peptides and proteins. These biologically active molecules can be used for evaluation of chronic as well as acute toxicity of a toxic metal or as biomarkers of assessment of environmental pollution (Fig. 2).

3.1 Physiological changes in sunflower plants exposed to silver(I) ions

To investigate influence of silver(I) ions on plant we selected sunflower plants (Helianthus annuus L., Asteraceae, syn. Compositae). The sunflower plants were exposed to silver(I) ions (0, 0.1, 0.5, and 1 mM) for 96 h. The images of the plants in time scale of the treatment are shown in Fig. 3. It clearly follows from the figures that the plants exposed to silver(I) ions embody growth depression; only control set of the plants demonstrates leaves growth. The root system of the plants exposed to silver(I) ions shows also physiological changes compared to control plants. Particularly we observed growth depression and coloured changes that indicate the entry of silver ions to the plants. Moreover the sunflower plants treated with silver(I) ions lack root hairs (Fig.3).

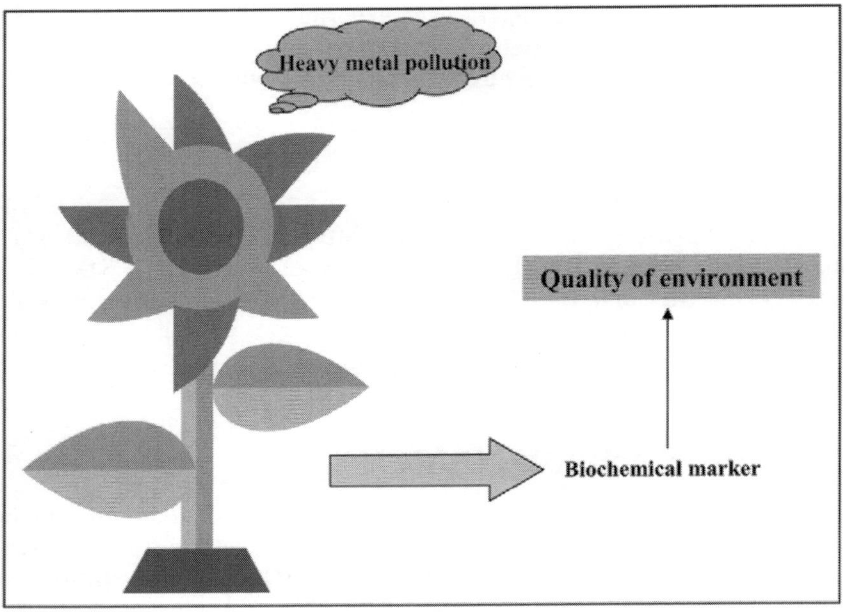

Figure 2. The scheme of environment quality assessment based on determination of biochemical markers presented in plants.

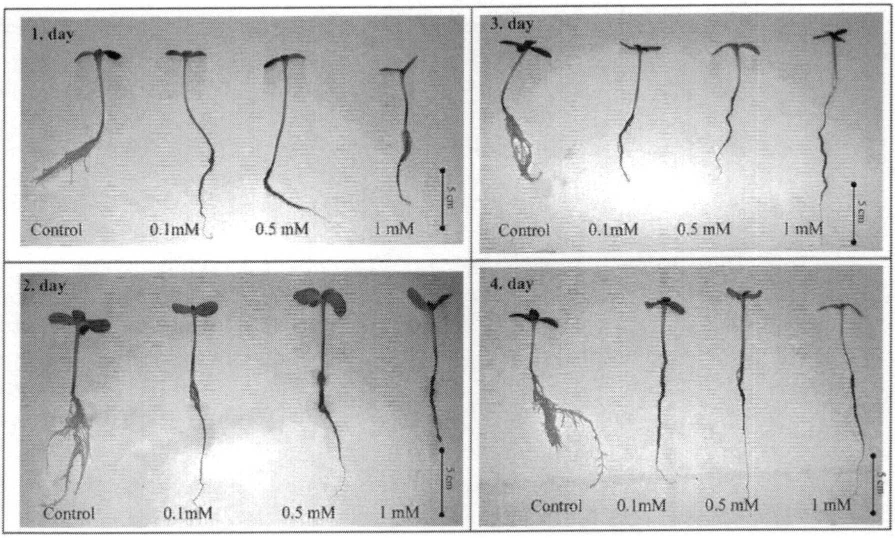

Figure 3. Images of the sunflower plants during 96 h treatment with silver(I) ions.

3.2 Changes in the treated plants fresh weight

At the end of each experimental day the part of the plants were harvested and weighed. Based on these results the growth curves were determined (Fig. 4). The control plants demonstrated moderate loss of fresh weight (Fig. 4 whole plant) because of their cultivation in distilled water without macro-and microelements supplement. This effect is mostly observable at shoots; root systems were more stimulated (Fig. 4). The influence of silver(I) ions on sunflower plants growth parameters was well evident. Rate of loss of plants fresh weight enhanced with the increasing silver(I) ions doses and with the time of the exposure. This effect was more evident at the shoot parts of the plants (hypocotyls, stems) in comparison with the root system. The decrease in the fresh weight is probably connected with an increase in a metabolic activity of the plants exposed to silver(I) ions due to very limited supply of inorganic as well as organic compounds needed for plant development. In addition we found that the content of chlorophylls (chlorophyll a and chlorophyll b) was lower in the plants treated with silver(I) ions compared to the control group of the plants. This indicates the damage of photosynthetic processes and the decrease of the total energetic metabolism of the plants treated with silver(I) ions.

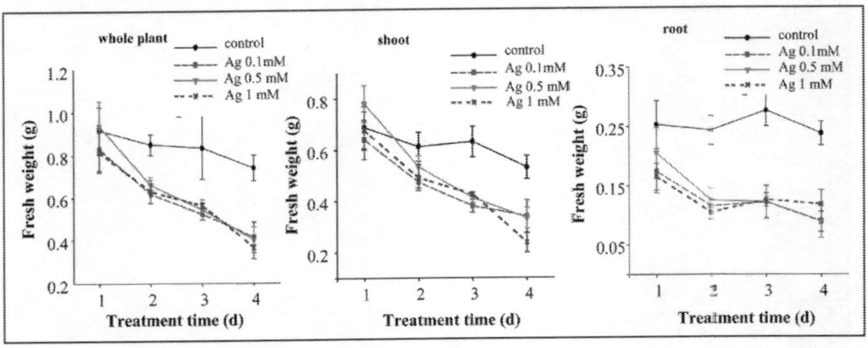

Figure 4. Fresh weight changes (whole plant, shoot and root) of sunflower plants exposed to various doses of silver(I) ions for 96 h.

3.3 Anatomical changes of sunflower shoots (hypocotyls, stems) and roots

Using of autofluorescence of anatomical structures, such as lignified cell walls, it was possible to determine the changes of important shoot and root structures, mainly vascular bungles and development of secondary thickening. The differences in vascular bundles organisation, parenchymatic pith development in the root centre and the reduction of phloem part of vascular bundles are well observable (Fig. 5). The most important changes in hypocotyls anatomy were as follows: the increased robustness of cuticle, increase of lignification and suberinization of the cell walls of epidermis and outer cortex, and changes in new vascular bundles differentiation with increasing silver(I) ions dose. These hallmarks can be related to changes of water transport caused by silver(I) ions.

Very evident and characteristic changes were determined in root structure. With increasing silver(I) ions concentration the vitality of rhizodermal cells declined; rhizodermal cells early necrosed and were replaced by the cells of exodermis. Changes were well evident in the structure of stele, mainly in pith development. Deposition of silver in the pitch cell walls as well as in inner xylem parts was expressively evident in connection with increasing silver(I) ions concentration.

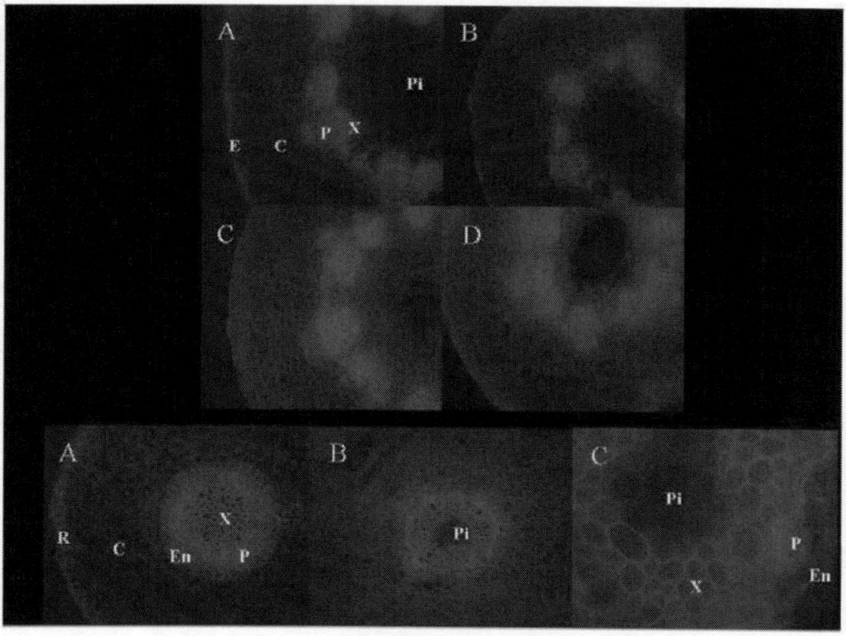

Figure 5. Sunflower shoot. Transverse stem section of sunflower cultivated without (**A**) and treated with silver(I) ions – 0.1 mM (**B**), 0.5 mM (**C**) and 1.0 mM (**D**) at fourth day of the treatment. Blue fluorescence – lignified cell walls. E – epidermis, C – cortex, P – phloem, X – xylem, Pi – pith. **Sunflower root.** Transverse root section of sunflower cultivated without (**A**) and treated with silver(I) ions – 0.1 mM (**B**), with detail (**C**) at fourth day of the treatment. Magnification × 100.

3.4 Spatial distribution of silver(I) ions in the plant tissues

Laser Induced Breakdown Spectroscopy (LIBS) is a simple spectrochemical sensor technology. LIBS utilises the emission of a high-temperature microplasma created by focusing a laser beam to the surface of the investigated sample. The spectrum of light emission is collected by a detector and its intensity at specific wavelengths is recorded [57]. The method allows real-time multi-element analysis in-situ and even remotely. The advantages of this method, i.e. usually none or very simple sample preparation, together with the capability of analysis with high spatial resolution [58] make it possible to utilize it in a wide-range of applications [57,58]. Basically, the achievable spatial resolution of LIBS is determined by the size of the laser craters produced within the material [57]. However, LIBS as a technique for simultaneous multi-elemental analysis is dating from the 1980s [59,60], as an emerging chemical sensor has picked up considerably during the past decade [57]. Recently, the LIBS was exploit among other for monitoring of accumulation of selected chemical elements in different structures of plant species [61-63].

The single-shot LIBS analysis was performed along the ~25 mm long Helianthus annuus L. stem sections. The LIBS's ICCD detector was gated 1 μs after the Q-switch signal and the observation window was 10 μs. Typical LIBS spectrum is shown in Fig. 6aA), together with a photograph of LIBS ablation craters (Fig. 6B). For Ag and Cu detection the 328.07 nm Ag(I) and 324.75 nm Cu(I) lines were used, respectively. For the mapping the background was subtracted (for each shot) and the area under the selected peak (for appropriate chemical element) was calculated.

Figure 6 demonstrates the capability of LIBS to signalize the trends in accumulation of different elements in selected structures of the sample. To trace the uptake of Ag and Cu in the investigated area of the leaf samples, a statistical analysis was carried out on the intensity data set from each sample. Because the measured intensity values had a log-normal distribution the average of the logarithm of intensity values was calculated [64]. From these data we found that in the leaves of the plants the overall Cu content follows the changes in the Ag content.

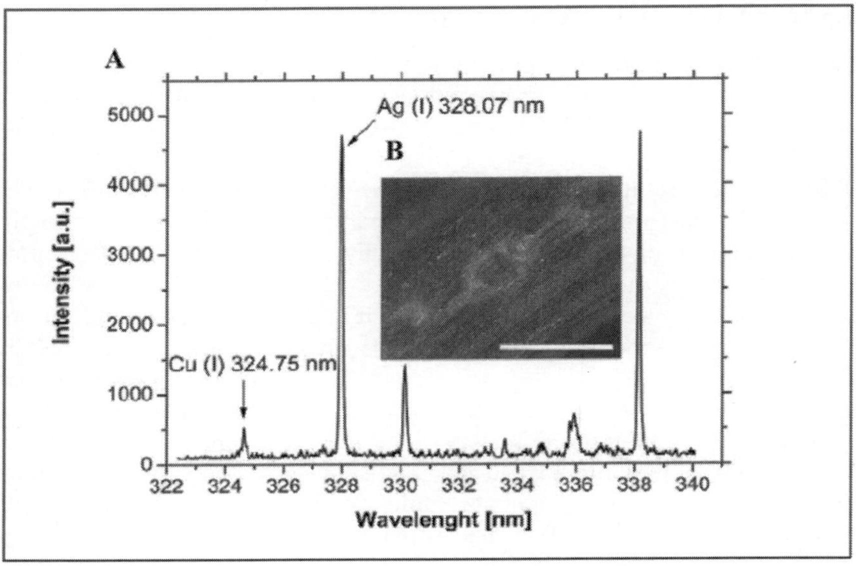

Figure 6. The typical LIBS spectrum with the 328.07 nm Ag(I) and 324.75 nm Cu(I) lines used in the analysis (**A**) together with the image of three ablation craters (**B**). The bar on **B** has a length of 500 μm.

The distribution of Cu and Ag along the stem sections for different days of treatment is shown in Figs. 7 and 8. The LIBS analysis was finished on the closer part of stem towards the root system (position ~25 mm on the sample). These graphs clearly demonstrate the advantage of the spatially-resolved LIBS analysis. While the overall content of the Ag and Cu follows each other (Fig. 7), the spatial distribution of these elements is different. The Ag is accumulated mainly in near-root part of the sample (Fig. 8A), on contrary the Cu is spread more uniformly within the stem (Fig. 8B).

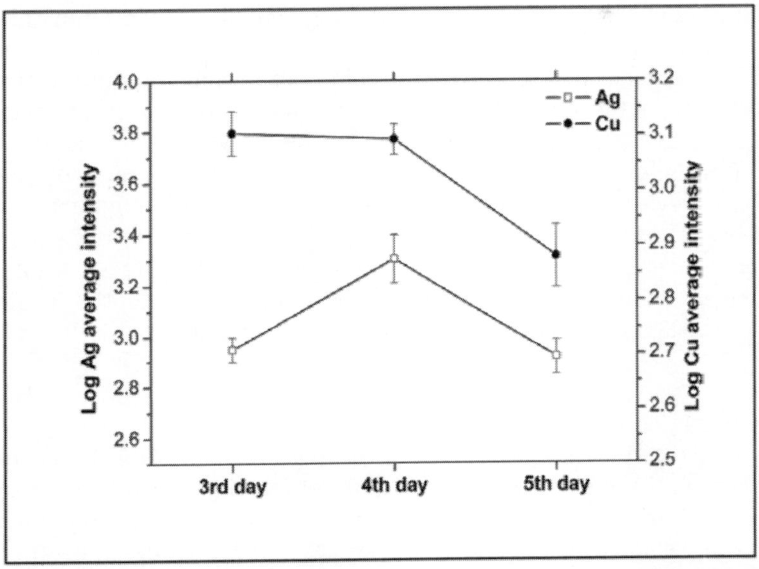

Figure 7. The average intensities calculated from logarithm of intensity data obtained from LIBS analyses.

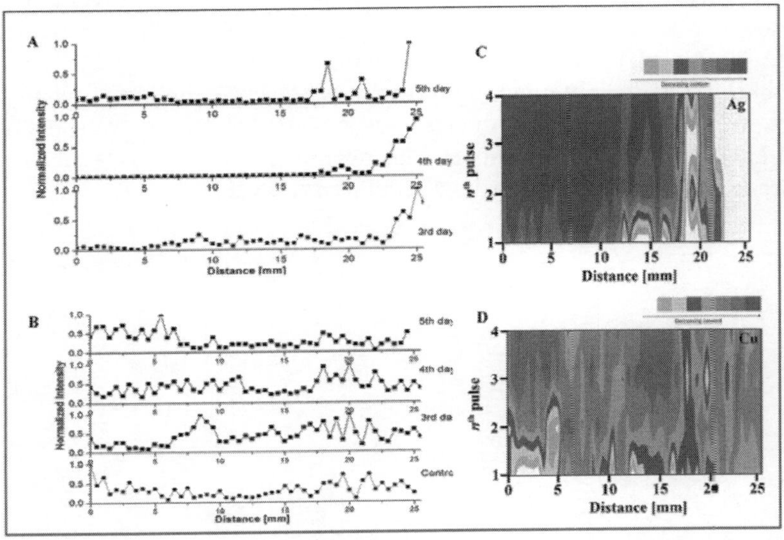

Figure 8. Accumulation of silver (**A**) and copper (**B**) along the investigated stem samples. Maps of the accumulated silver (**C**) and copper (**D**) measured by the LIBS technique for quasy 3D analyses of the stem sample of interest.

The possibility to utilize LIBS for 3D analysis of element distribution within the sample is examined in Fig. 8C,D. The different layers were monitored with four subsequent laser pulses. We should note that in spite to the fact that on the frame of this exploratory study the depth of ablation craters was not measured, the different distribution of elements within the analyzed layers is clearly observable. On the ongoing work an upgrade of the instrumental device for simultaneous LIBS – laser ablation inductively coupled plasma mass spectrometry (LA-ICP-MS) is planned. Utilizing such an apparatus, also the calibration for the main components will be possible, using selected calibration standards and measuring the depth of the ablation-craters.3.5 Biochemical indicators of environmental stress.

3.5.1 The total protein content

At the plants treated with silver(I) ions the changes were observable also in total content of proteins determined using biuret reaction. For this purpose automatic analyzer BS-200 was used. The calibration curve obtained (albumin was selected as standard) was strictly linear within the range from 1 to 1000 mg/ml (Fig. 9A). The level of water soluble proteins was markedly lower in root system, where also the difference between control plants and plants treated with silver(I) ions was mostly distinctive. In addition the total content of proteins expressively decreased with increasing silver(I) ions dose and the time of the treatment (Fig. 9B). Shoots of sunflower plants demonstrated less distinct proteins content in plants treated with silver(I) ions in comparison with root system (Fig. 9C). As we compare the results obtained by protein analysis – the total protein contents in shoot as well as root parts – we can assume on the transport of the proteins from the roots to shoots. This phenomenon can be related with the cascade of processes connecting with photosynthesis.

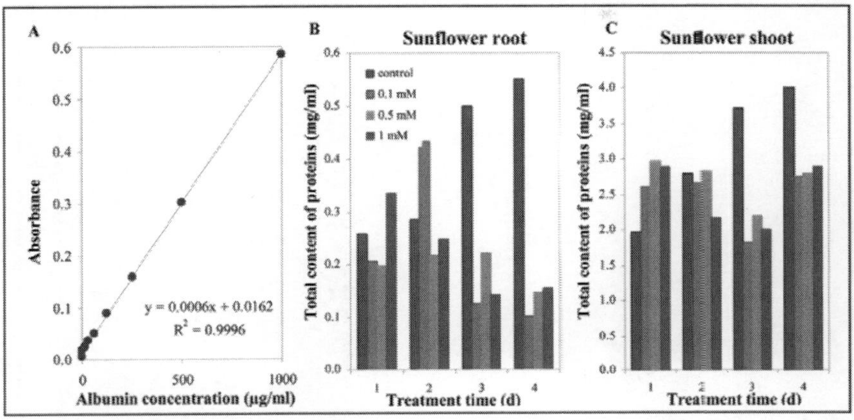

Figure 9. Total content of proteins soluble in water determined spectrometrically using biuret reaction. The dependence of signal height on albumin concentration (**A**). Total content of proteins in root (**B**) and shoots (**C**) of sunflower plants exposed to silver(I) ions.

3.5.2 Urease activity

One of the crucial plant enzymes, which are extremely sensitive to presence of heavy metals ions, is urease. Due to this feature the number of different biosensors to detect heavy metals was proposed [65,66]. All experiments appear from enzymes that are very carefully purified. However investigation of enzyme activities for environment pollution assessment is also interesting. Sunflower plants demonstrate measurable level of urease activity. Because of this fact we aimed on the monitoring of urease activity. For our purposes the methodology enabling urease activity determination using automated detection was optimised. Automated spectrophotometric analysator BS-200 was used. The principle of this method is based on the fact that urease decomposes carbamide into ammonia that is subsequently detectable as coloured product [65,67]. The strictly linear calibration curve is shown in Fig. 10A. Relative standard deviation was about 1.5 % and the analysis of 40 samples was realized within 30 min. For the examination of the activity of enzyme urease under our experimental conditions enzyme kinetic was observed. The measurements resulted in Michaelis-Menten constant Km 3.8 µM (Fig. 10B).

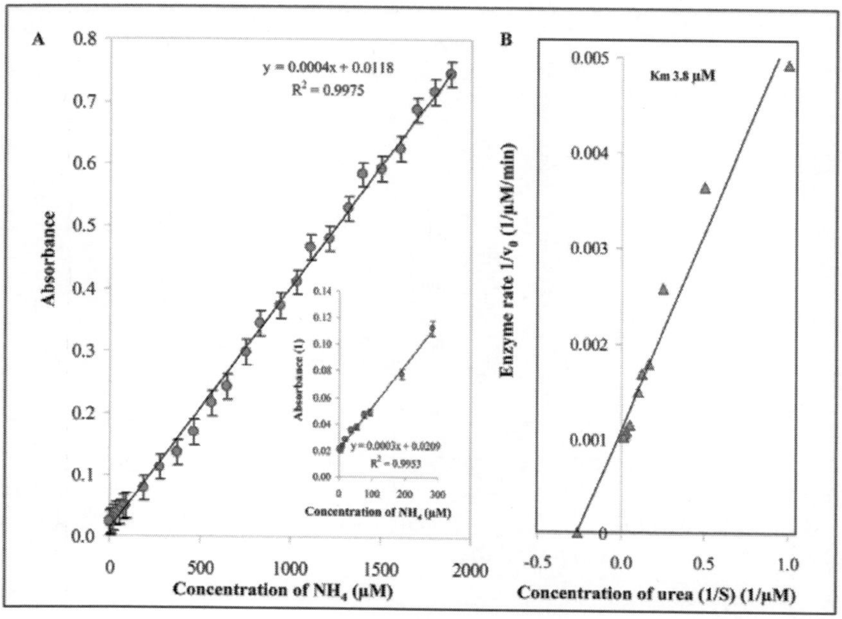

Figure 10. Spectrometric automated analysis of urease activity. The dependence of total amount of released ammonia on height of the obtained signal (**A**). Enzyme kinetic of soybean urease (**B**).

Sunflower plants demonstrated higher urease activity in root system in comparison with shoots (Fig. 11A,B). The activity in roots was app. five times higher compared to shoots. If we compared the activity in treated plants with control, we found out that presence of silver(I) ions markedly enhanced the activity of urease at all applied doses of this toxic metal.

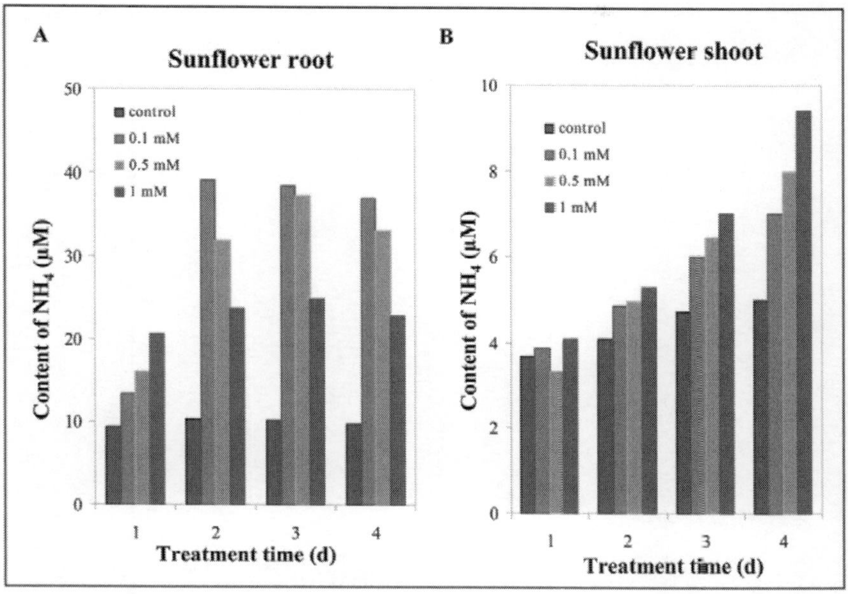

Figure 11. The changes in urease activity in sunflower plants exposed to silver(I) ions, roots (**A**) and shoots (**B**).

3.6 The changes of urease activity in the presence of silver(I) ions

In addition we were interested in the issue how can the presence of silver(I) ions influence the activity of purified urease. Into the solution of urea (1 mM) silver(I) ions at final concentrations 0; 5; 10; 15; 20; 25; 30; 35; 40; 60; 80 and/or 100 µM were added. To this solution urease (10 IU) was added and mixture was incubated for 60 s at 37 °C under shaking (300 rpm). The samples were analysed using procedure described above. Urease activity was inhibited by silver(I) ions; concentrations higher than 20 mM quickly led to reduction in urease activity (Fig. 12)

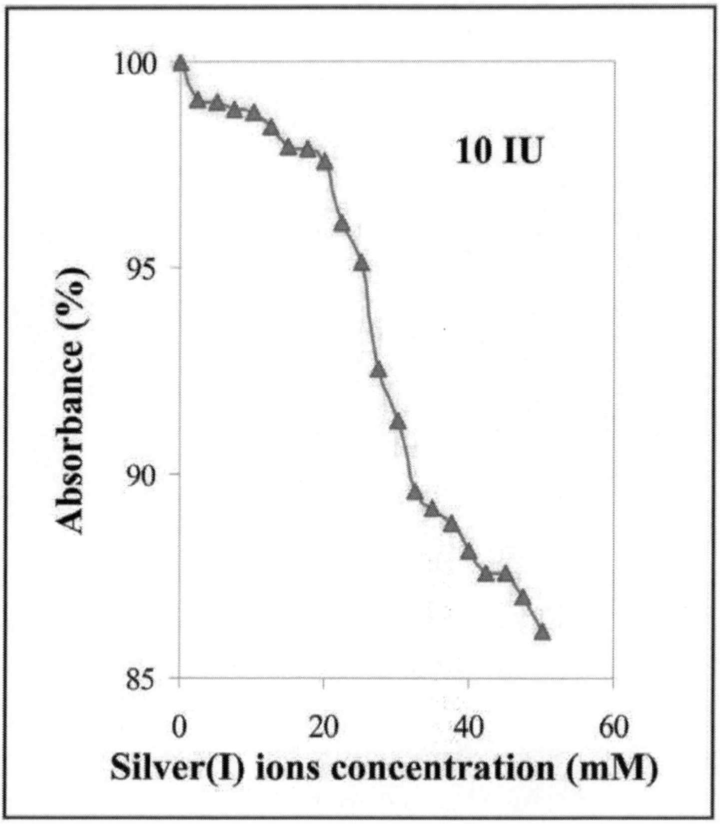

Figure 12. The changes of activity of commercially available urease in the presence of silver(I) ions.

4. CONCLUSION

The plants would be used as simple bioindicators of the quality of environment [68-73]. However it is absolutely necessary to understand some physiological processes that are connected with the plant response on stress. The data obtained are, despite of this fact, very encouraging to perform deeper research in this area.

ACKNOWLEDGMENTS

We gratefully acknowledge the Grant Agency of the Czech Republic (grant No. GA CR 526/07/0674) for financial support to this work.

REFERENCES

1. Galvez, F.; Hogstrand, C.; McGeer, J.C.; Wood, C.M. The physiological effects of a biologically incorporated silver diet on rainbow trout (Oncorhynchus mykiss). Aquat. Toxicol. **2001**, 55, 95–112
2. Davis, A.; Drexler, J.W.; Ruby, M.V.; Nicholson, A. Micromineralogy of Mine Wastes in Relation to Lead Bioavailability, Butte, Montana. Environ. Sci. Technol. **1993**, 27, 1415–1425.
3. Hogstrand, C.; Wood, C.M. Toward a better understanding of the bioavailability, physiology and toxicity of silver in fish: Implications for water quality criteria. Environ. Toxicol. Chem. **1998**, 17, 547–561.
4. Mann, R.M.; Ernste, M.J.; Bell, R.A.; Kramer, J.R.; Wood, C.M. Evaluation of the protective effects of reactive sulfide on the acute toxicity of silver to rainbow trout (Oncorhynchus mykiss). Environ. Toxicol. Chem. **2004**, 23, 1204–1210.
5. Hogstrand, C.; Wood, C.M. Toward a better understanding of the bioavailability, physiology and toxicity of silver in fish: Implications for water quality criteria. Environ. Toxicol. Chem. **1998**, 17, 547–561.
6. Gorsuch, J.W.; Klaine, S.J. Toxicity and fate of silver in the environment. Environ. Toxicol. Chem. **1998**, 17, 537–538.
7. Song, Y.M.; Lu, X.L.; Yang, M.L.; Zheng, X.R. Study on the interaction of platinum(IV), gold(III) and silver(I) ions with DNA. Transit. Met. Chem. **2005**, 30, 499–502.
8. Hossain, Z.; Huq, F. Studies on the interaction between Ag– and DNA. J. Inorg. Biochem. **2002**, 91, 398–404.
9. Clemens, S. Toxic metal accumulation, responses to exposure and mechanisms of tolerance in plants. Biochimie **2006**, 88, 1707–1719.
10. Zenk, M.H. Heavy metal detoxification in higher plants - A review. Gene **1996**, 179, 21–30. [Google Scholar]
11. Hamer, D.H. Metallothionein. Annu. Rev. Biochem. **1986**, 55, 913–951.
12. Giordani, T.; Natali, L.; Maserti, B.E.; Taddei, S.; Cavallini, A. Characterization and expression of DNA sequences encoding putative

type-II metallothioneins in the seagrass Posidonia oceanica. Plant Physiol. **2000**, 123, 1571–1581.
13. Sochova, I.; Hofman, J.; Holoubek, I. Using nematodes in soil ecotoxicology. Environ. Int. **2006**, 32, 374–383.
14. Fichez, R.; Adjeroud, M.; Bozec, Y.M.; Breau, L.; Chancerelle, Y.; Chevillon, C.; Douillet, P.; Fernandez, J.M.; Frouin, P.; Kulbicki, M.; Moreton, B.; Ouillon, S.; Payri, C.; Perez, T.; Sasal, P.; Thebault, J. A review of selected indicators of particle, nutrient and metal inputs in coral reef lagoon systems. Aquat. Living Resour. **2005**, 18, 125–147.
15. Bebianno, M.J.; Geret, F.; Hoarau, P.; Serafim, M.A.; Coelho, M.R.; Gnassia-Barelli, M.; Romeo, M. Biomarkers in Ruditapes decussatus: a potential bioindicator species. Biomarkers **2004**, 9, 305–330.
16. Conti, M.E.; Cecchetti, G. Biological monitoring: lichens as bioindicators of air pollution assessment - a review. Environ. Pollut. **2001**, 114, 471–492.
17. McGeoch, M.A. The selection, testing and application of terrestrial insects as bioindicators. Biol. Rev. Cambridge Philosophic. Soc. **1998**, 73, 181–201.
18. Leyval, C.; Turnau, K.; Haselwandter, K. Effect of heavy metal pollution on mycorrhizal colonization and function: physiological, ecological and applied aspects. Mycorrhiza **1997**, 7, 139–153.
19. Petrlova, J.; Krizkova, S.; Zitka, O.; Hubalek, J.; Prusa, R.; Adam, V.; Wang, J.; Beklova, M.; Sures, B.; Kizek, R. Utilizing a chronopotentiometric sensor technique for metallothionein determination in fish tissues and their host parasites. Sens. Actuator B-Chem. **2007**, 127, 112–119.
20. Lessire, F.; Delaunois, A.; Gustin, P.; Ansay, M. Biomarkers and bioindicators in vertebrates: importance in evaluation of quality of an ecosystem. Ann. Med. Vet. **1997**, 141, 281–290.
21. Kafka, Z.; Puncocharova, J. Bioindicators in the environment monitoring. Chem. Listy **2000**, 94, 909–912.
22. Wyrobek, A.J.; Schmid, T.E.; Marchetti, F. Cross-species sperm-FISH assays for chemical testing and assessing paternal risk for chromosomally abnormal pregnancies. Environ. Mol. Mutagen. **2005**, 45, 271–283.
23. Sures, B. Environmental parasitology: relevancy of parasites in monitoring environmental pollution. Trends Parasitol. **2004**, 20, 170–177.

24. Purcell, T.W.; Peters, J.J. Sources of silver in the environment. Environ. Toxicol. Chem. **1998**, 17, 539–546.
25. Purcell, T.W.; Peters, J.J. Historical impacts of environmental regulation of silver. Environ. Toxicol. Chem. **1999**, 18, 3–8.
26. Saeki, S.; Kubota, M.; Asami, T. Determination of silver in plants by flame atomic absorption spectrometry. Int. J. Environ. Anal. Chem. **1996**, 64, 179–183.
27. Shamspur, T.; Mashhadizadeh, M.H.; Sheikhshoaie, I. Flame atomic absorption spectrometric determination of silver ion after preconcentration on octadecyl silica membrane disk modified with bis[5-((4-nitrophenyl)azosalicylaldehyde)] as a new Schiff base ligand. J. Anal. At. Spectrom. **2003**, 18, 1407–1410.
28. Raoof, J.B.; Ojani, R.; Kiani, A. Kinetic determination of silver ion by its perturbation on Belousov-Zhabotinskii oscillating chemical reaction using potentiometric method. Anal. Sci. **2004**, 20, 883–886.
29. Safavi, A.; Iranpoor, N.; Saghir, N. Directly silica bonded analytical reagents: synthesis of 2-mercaptobenzothiazole-silica gel and its application as a new sorbent for preconcentration and determination of silver ion using solid-phase extraction method. Sep. Purif. Technol. **2004**, 40, 303–308.
30. Schildkraut, D.E.; Dao, P.T.; Twist, J.P.; Davis, A.T.; Robillard, K.A. Determination of silver ions at sub microgram-per-liter levels using anodic square-wave stripping voltammetry. Environ. Toxicol. Chem. **1998**, 17, 642–649.
31. Zhang, S.B.; Zhang, X.J.; Lin, X.Q. An ethylenediaminetetraacetic acid modified carbon paste electrode for the determination of silver ion. Chin. J. Anal. Chem. **2002**, 30, 745–747.
32. Guo, S.X.; Khoo, S.B. Highly selective and sensitive determination of silver(I) at a poly(8-mercaptoquinoline) film modified glassy carbon electrode. Electroanalysis **1999**, 11, 891–898.
33. Ye, X.Z.; Yang, Q.H.; Wang, Y.; Li, N.Q. Electrochemical behaviour of gold, silver, platinum and palladium on the glassy carbon electrode modified by chitosan and its application. Talanta **1998**, 47, 1099–1106.
34. Wang, J.; Lu, J.M.; Farias, P.A.M. Remote electrochemical monitoring of trace silver. Anal. Chim. Acta **1996**, 318, 151–157.
35. Mikelova, R.; Baloun, J.; Petrlova, J.; Adam, V.; Havel, L.; Petrek, J.; Horna, A.; Kizek, R. Electrochemical determination of Ag-ions in environment

waters and their action on plant embryos. Bioelectrochemistry **2007**, 70, 508–518.
36. Stejskal, K.; Krizkova, S.; Adam, V.; Sures, B.; Trnkova, L.; Zehnalek, J.; Hubalek, J.; Beklova, M.; Hanustiak, P.; Svobodova, Z.; Horna, A.; Kizek, R. Bio-assessing of environmental pollution via monitoring of metallothionein level using electrochemical detection. IEEE Sens. J. **2007**. in press.
37. Svancara, I.; Ogorevc, B.; Hocevar, S.B.; Vytras, K. Perspectives of carbon paste electrodes in stripping potentiometry. Anal. Sci. **2002**, 18, 301–305.
38. Svancara, I.; Ogorevc, B.; Novic, M.; Vytras, K. Simple and rapid determination of iodide in table salt by stripping potentiometry at a carbon-paste electrode. Anal. Bioanal. Chem. **2002**, 372, 795–800.
39. Barek, J.; Muck, A.; Wang, J.; Zima, J. Study of voltammetric determination of carcinogenic 1-nitropyrene and 1-aminopyrene using a glassy carbon paste electrode. Sensors **2004**, 4, 47–57.
40. Svancara, I.; Kalcher, K.; Diewald, W.; Vytras, K. Voltammetric determination of silver at ultratrace levels using a carbon paste electrode with improved surface characteristics. Electroanalysis **1996**, 8, 336–342.
41. Svancara, I.; Vytras, K.; Barek, J.; Zima, J. Carbon paste electrodes in modern electroanalysis. Crit. Rev. Anal. Chem. **2001**, 31, 311–345.
42. Lawrence, N.S.; Deo, R.P.; Wang, J. Biocatalytic carbon paste sensors based on a mediator pasting liquid. Anal. Chem. **2004**, 76, 3735–3739.
43. Blaedel, W.J.; Wang, J. Mixed Immobilized Enzyme-Porous Electrode Reactor. Anal. Chem. **1980**, 52, 1426–1429.
44. Kizek, R.; Vacek, J.; Trnkova, L.; Klejdus, B.; Kuban, V. Electrochemical biosensors in agricultural and environmental analysis. Chem. Listy **2003**, 97, 1003–1006.
45. Masarik, M.; Kizek, R.; Kramer, K.J.; Billova, S.; Brazdova, M.; Vacek, J.; Bailey, M.; Jelen, F.; Howard, J.A. Application of avidin-biotin technology and adsorptive transfer stripping square-wave voltammetry for detection of DNA hybridization and avidin in transgenic avidin maize. Anal. Chem. **2003**, 75, 2663–2669.
46. Petrlova, J.; Masarik, M.; Potesil, D.; Adam, V.; Trnkova, L.; Kizek, R. Zeptomole detection of streptavidin using carbon paste electrode and square wave voltammetry. Electroanalysis **2007**, 19, 1177–1182.

47. Berry, W.J.; Cantwell, M.G.; Edwards, P.A.; Serbst, J.R.; Hansen, D.J. Predicting toxicity of sediments spiked with silver. Environ. Toxicol. Chem. **1999**, 18, 40–48.
48. Bury, N.R.; McGeer, J.C.; Wood, C.M. Effects of altering freshwater chemistry on physiological responses of rainbow trout to silver exposure. Environ. Toxicol. Chem. **1999**, 18, 49–55.
49. Bury, N.R.; Galvez, F.; Wood, C.M. Effects of chloride, calcium, and dissolved organic carbon on silver toxicity: Comparison between rainbow trout and fathead minnows. Environ. Toxicol. Chem. **1999**, 18, 56–62.
50. Bianchini, A.; Wood, C.M. Mechanism of acute silver toxicity in Daphnia magna. Environ. Toxicol. Chem. **2003**, 22, 1361–1367.
51. Bianchini, A.; Bowles, K.C.; Brauner, C.J.; Gorsuch, J.W.; Kramer, J.R.; Wood, C.M. Evaluation of the effect of reactive sulfide on the acute toxicity of silver (I) to Daphnia magna. part 2: Toxicity results. Environ. Toxicol. Chem. **2002**, 21, 1294–1300.
52. Bianchini, A.; Grosell, M.; Gregory, S.M.; Wood, C.M. Acute silver toxicity in aquatic animals is a function of sodium uptake rate. Environ. Sci. Technol. **2002**, 36, 1763–1766.
53. Call, D.J.; Polkinghorne, C.N.; Markee, T.P.; Brooke, L.T.; Geiger, D.L.; Gorsuch, J.W.; Robillard, K.A. Silver toxicity to Chironomus tentans in two freshwater sediments. Environ. Toxicol. Chem. **1999**, 18, 30–39.
54. Karen, D.J.; Ownby, D.R.; Forsythe, B.L.; Bills, T.P.; La Point, T.W.; Cobb, G.B.; Klaine, S.J. Influence of water quality on silver toxicity to rainbow trout (Oncorhynchus mykiss), fathead minnows (Pimephales promelas), and water fleas (Daphnia magna). Environ. Toxicol. Chem. **1999**, 18, 63–70.
55. Morgan, T.P.; Wood, C.M. A relationship between gill silver accumulation and acute silver toxicity in the freshwater rainbow trout: Support for the acute silver biotic ligand model. Environ. Toxicol. Chem. **2004**, 23, 1261–1267.
56. Ratte, H.T. Bioaccumulation and toxicity of silver compounds: A review. Environ. Toxicol. Chem. **1999**, 18, 89–108. [Google Scholar]
57. Miziolek, A.W.; Palleschi, V.; Schecher, I. Laser-Induced Breakdown Spectroscopy (LIBS); Cambridge University Press, 2006.
58. Buckley, S.G. LIBS comes on strong. Laser Focus World **2006**, 42, 95–98.

59. Radziemski, L.J.; Loree, T.R.; Cremers, D.A.; Hoffman, N.M. Time-Resolved Laser-Induced Breakdown Spectrometry of Aerosols. Anal. Chem. **1983**, 55, 1246–1252.
60. Cremers, D.A.; Radziemski, L.J. Detection of Chlorine and Fluorine in Air by Laser-Induced Breakdown Spectrometry. Anal. Chem. **1983**, 55, 1252–1256.
61. Martin, M.Z.; Wullschleger, S.D.; Garten, C.T.; Palumbo, A.V.; Smith, J.G. Elemental analysis of environmental and biological samples using laser-induced breakdown spectroscopy and pulsed Raman spectroscopy. J. Dispersion Sci. Technol. **2004**, 25, 687–694.
62. Samek, O.; Lambert, J.; Hergenroder, R.; Liska, M.; Kaiser, J.; Novotny, K.; Kukhlevsky, S. Femtosecond laser spectrochemical analysis of plant samples. Laser Phys. Lett. **2006**, 3, 21–25.
63. Kaiser, J.; Samek, O.; Reale, L.; Liska, M.; Malina, R.; Ritucci, A.; Poma, A.; Tucci, A.; Flora, F.; Lai, A.; Mancini, L.; Tromba, G.; Zanini, F.; Faenov, A.; Pikuz, T.; Cinque, G. Monitoring of the heavy-metal hyperaccumulation in vegetal tissues by X-ray radiography and by femto-second laser induced breakdown spectroscopy. Microsc. Res. Tech. **2007**, 70, 147–153.
64. Limpert, E.; Stahel, W.A.; Abbt, M. Log-normal distributions across the sciences: Keys and clues. Bioscience **2001**, 51, 341–352.
65. Hubalek, J.; Hradecky, J.; Adam, V.; Krystofova, O.; Huska, D.; Masarik, M.; Trnkova, L.; Horna, A.; Klosova, K.; Adamek, M.; Zehnalek, J.; Kizek, R. Spectrometric and voltammetric analysis of urease – Nickel nanoelectrode as an electrochemical sensor. Sensors **2007**, 7, 1238–1255.
66. Xu, Z.; Chen, X.; Qu, X.H.; Jia, J.B.; Dong, S.J. Single-wall carbon nanotube-based voltammetric sensor and biosensor. Biosens. Bioelectron. **2004**, 20, 579–584.
67. Witte, C.P.; Medina-Escobar, N. In-gel detection of urease with nitroblue tetrazolium and quantification of the enzyme from different crop plants using the indophenol reaction. Anal. Biochem. **2001**, 290, 102–107.
68. Petrek, J.; Vitecek, J.; Vlasinova, H.; Kizek, R.; Kramer, K.J.; Adam, V.; Klejdus, B.; Havel, L. Application of computer imaging, stripping voltammetry and mass spectrometry for study of the effect of lead (Pb-EDTA) on growth and viability of early somatic embryos of Norway spruce (Picea abies /L./ Karst.). Anal. Bioanal. Chem. **2005**, 383, 576–586.

69. Supalkova, V.; Beklova, M.; Baloun, J.; Singer, C.; Sures, B.; Adam, V.; Huska, D.; Pikula, J.; Rauscherova, L.; Havel, L.; Zehnalek, J.; Kizek, R. Affecting of aquatic vascular plant Lemna minor by cisplatin revealed by voltammetry. Bioelectrochemistry **2007**. in press.
70. Supalkova, V.; Huska, D.; Diopan, V.; Hanustiak, P.; Zitka, O.; Stejskal, K.; Baloun, J.; Pikula, J.; Havel, L.; Zehnalek, J.; Adam, V.; Trnkova, L.; Beklova, M.; Kizek, R. Electroanalysis of plant thiols. Sensors **2007**, 7, 932–959.
71. Supalkova, V.; Petrek, J.; Baloun, J.; Adam, V.; Bartusek, K.; Trnkova, L.; Beklova, M.; Diopan, V.; Havel, L.; Kizek, R. Multi-instrumental investigation of affecting of early somatic embryos of Spruce by cadmium(II) and lead(II) ions. Sensors **2007**, 7, 743–759.
72. Vitecek, J.; Adam, V.; Petrek, J.; Vacek, J.; Kizek, R.; Havel, L. Esterases as a marker for the growth of BY-2 tobacco cells and early somatic embryos of the norway spruce. Plant. Cell. Tiss. Org. **2004**, 79, 195–201.
73. Zitka, O.; Stejskal, K.; Kleckerova, A.; Adam, V.; Beklova, M.; Horna, A.; Havel, L.; Kizek, R. Utilizing of electrochemical techniques for detection of biological samples. Chem. Listy **2007**, 101, 225–231.

10

CHAPTER

BANANA LECTIN: A BRIEF REVIEW

Senjam Sunil Singh [1], Sanjenbam Kunjeshwori Devi [1,*] and Tzi Bun Ng [2,*]

[1]*Laboratory of Protein Biochemistry, Biochemistry Department, Manipur University, Canchipur, Imphal 795003, India*

[2]*School of Biomedical Sciences, Faculty of Medicine, The Chinese University of Hong Kong, Shatin, Hong Kong, China*

ABSTRACT

Lectins are a group of proteins of non-immune origin that recognize and bind to carbohydrates without modifying them. Banana is the common name for both herbaceous plants of the genus Musa and for the fruit they produce. They are indeed a promising source for many medicinal applications. Banana lectins have the potential for inhibiting HIV-1 reverse transcriptase activity, suppressing cancer cell proliferation and stimulating macrophage activities. Nevertheless, compared to other plant lectins, there is relatively little information in the literature on banana lectins, particularly with respect to their structure and biological functions. Herein we focus our review on the structure, functions and exploitable properties of banana lectins.

Keywords: Musa; banana lectins; reverse transcriptase; cancer cell proliferation; macrophage

1. INTRODUCTION

Banana is a major tropical fruit crop distributed in more than 120 countries with an annual production of 102 million tonnes. Banana belongs to the family Musaceae under the order of Zingiberales with two genera, Musa L. and Ensete Bruce [1]. It is highly diversified throughout the world but has been reported to originate from Southeast Asia [2]. The genus Musa consists of around 50 species while Ensete has nine species [3]. On the basis of phenotypic traits and basic chromosome number, Musa has been divided into four sections namely, Eumusa, Rhodochlamys, Austra-limusa and Callimusa. Most edible bananas that have largely evolved from two wild species, M. acuminata Colla and M. balbisiana Colla belong to the section Eumusa [4,5].

Traditionally banana has been used for numerous medicinal purposes. For example, banana can be useful in many forms for the treatment of diarrhea, and gastric ulcers, etc. Banana is an excellent source of potassium and a number of vitamins [6]. Jain et al. [7] investigated the antibacterial and antioxidant activities of different parts of seeded banana fruits in vitro. Dried peels, pulps and seeds of the fruit were extracted with hexane, ethyl acetate and ethanol. Antibacterial property of the extracts was evaluated against Gram-positive and Gram-negative bacteria using the disc diffusion technique. The antioxidant activities were determined by the 2,2-diphenyl-1-picrylhydrazyl (DPPH) scavenging system, ferric reducing ability in plasma (FRAP) assays and total phenolic content (TPC) assays [8]. It was shown that the extracts of local seedy banana fruits possessed significant antibacterial and antioxidant activities. Sampath et al. [6] also reported that all parts of the banana plant have medicinal applications and at the same time banana lectins exhibit the potential of inhibiting HIV-1 reverse transcriptase activity, suppressing cancer cell proliferation and stimulating macrophage activities. Nevertheless, compared to other plant lectins, very few banana lectins have been structurally characterized or produced in a recombinant form. Therefore the aim of this article is to provide an overview of the structure, functions and exploitable properties of banana lectins.

2. BANLEC

Lectins are a group of proteins of non-immune origin that recognize and bind to carbohydrates without modifying them [9]. Lectins are known to be important for many biological processes, due to their ability to recognize cell-surface carbohydrates with high specificity. Plant lectins have been model systems for studying protein-carbohydrate recognition, because individually they exhibit high sensitivity and as a group large diversity in recognizing carbohydrate structures. Some of the well-characterized roles of lectins are in cell-cell communication, cancer metastasis, embryogenesis, host-pathogen interactions, and tissue development [10]. A large-scale data acquisition and extensive analysis of sequences and structures of beta-prism-I or jacalin-related lectins (JRLs) have shown that hypervariability in the binding site loops generates carbohydrate recognition diversity [11].

During an investigation of the human antibody response to various foods, Koshte et al. [12] observed marked binding of IgG4 to banana (Musa paradisiaca) extract from its fruit. In view of the mucilaginous nature of the banana extract, they considered the possibility of non-specific binding to IgG4 and decided to purify the active principle. They found that this binding activity was largely removed by gel filtration on a Sephadex G-75 column and was recovered by elution with sugar (mannose). This clearly indicated the presence of a lectin. They named it as BanLec-I [13].

That was how the banana lectin (BanLec or BL) was first isolated from Musa paradisiaca by Koshte and colleagues [12]. They also reported that BanLec is a homodimeric protein that binds mannose and mannose-containing oligosaccharides and functions as a potent T-cell mitogen [9,13]. Later on, Peumans and colleagues cloned and sequenced BanLec (Musa acuminata) and confirmed that it is composed of two identical 15-kDa subunits demonstrating specificity toward mannose. Based on sequence homology and secondary structure predictions, they classified BanLec as a member of JRL [14].

Peumans et al. [14] identified one of the predominant proteins in the pulp of ripe bananas (Musa acuminata L.) and plantains (Musa spp.) as a lectin. The identification of these lectins in banana plants and plantains demonstrates that JRLs also occur in monocot species. This point clearly suggests that lectins are actually more widespread among higher plants than is believed.

3. ISOLATION

The banana and plantain lectins are the first documented examples of JRLs, which are present in abundance in the pulp and roots of mature fruits, but are apparently absent from other tissues like peel, leaf and corm. However, Clendennen et al. [15] demonstrated that after treatment of intact banana plants with methyl jasmonate, BanLec can be induced in leaves. This clearly indicates that lectin might be expressed as a response to abiotic or biotic stress factors.

The banana and plantain agglutinins (called BanLec and PlanLec, respectively) have been purified in sizeable quantities using a novel isolation procedure, such as by preventing adsorption of the lectins onto insoluble endogenous polysaccharides. Original isolation procedures for BanLec adopted traditional affinity chromatography and acetic acid extraction of overripe bananas, presumably because much of the starch has been degraded and the low pH further reduced starch–lectin interaction. However, Wearne et al. [16], reported the added disadvantages of this procedure which include a low yield (10 mg/kg) of purified lectin and it may be due to low solubility of the lectin in an acidic medium, and/or to degradation of lectin during the ripening process.

At the other extreme, one-step purification of lectins using sugar-immobilized gold nanoparticles (SGNPs) was developed by Nakamura et al. [17]. With the use of such glycan-coated nanoparticles, a nearly 100-fold greater yield (1 μg/mg) was achieved. They used the 60% ammonium sulfate fraction obtained from the crude plant extract, incubated it with SGNPs overnight at 4 °C to allow to form aggregate of lectin and SGNPs. The lectin was finally dissolved by adding inhibitory sugars. These sugars are similar or identical to the non-reducing sugar moieties on the SGNPs. This procedure reflects the ability of the nanoparticles, coated with a high density of reactive mannosyl/glucosyl groups, to successfully compete with the starch of banana pulp for binding with the lectin [16,17].

4. STRUCTURES

Plant lectins have been classified into five structural families [18] in terms of their polypeptide folds. Of these, the β-prism I fold in lectins was first characterized by Sankaranarayanan et al. in jacalin from jackfruit seeds [19]. The other lectin from jackfruit seeds with different sugar specificity, artocarpin, also exhibits the same fold [18]. Originally, the β-prism I fold was thought to be characteristic of Moraceae lectins. However, lectins from other families were subsequently shown using X-ray analysis to display the same fold: e.g., calsepa from Calystegia sepium (family Convolvulacea) and heltuba from Helianthus tuberosus (family Asteraceae) [18]. On the basis of sequence similarity, Peumans et al., have also disclosed that banana lectin also manifests the same β-prism I fold [14].

Peumans et al. [14], showed that both BanLec (banana lectin) and PlanLec (plantain lectin) are dimeric proteins composed of two identical 15-kDa subunits each of which contains 141 amino acid residues. By molecular cloning and molecular modelling studies of the protein they also revealed that BanLec has sequence similarity to previously described lectins of the family of JRLs, possessing the same overall fold and three-dimensional structure [9]. Raval et al. [11], classified the JRLs family into three categories based on their quaternary structures. The first group contains jacalin and artocarpin, the second group contains calsepa and the third group contains heltuba [19]. Meagher et al. [9], proposed the fourth structural category of JRLs which contains the BanLec because its quaternary structure shows a unique dimer [9]. Dimeric banana lectin and calsepa, tetrameric artocarpin and octameric heltuba are mannose-specific β-prism I fold lectins with nearly the same tertiary structure [20].

Khan et al. [21], proposed the ribbon-shaped structure in BanLec (homodimeric protein), in which each subunit has a single Trp residue at position 10. Also each subunit has twelve β strands arranged in a β-prism-I fold. More importantly, the structure of BanLec obtained by X-ray crystallography revealed the presence of a second sugar binding site. The residues involved in the second site are common to other lectins in this family, potentially signaling a new group of mannose-specific jacalin-related lectins (mJRL) with two sugar binding sites i.e., two primary sugars-binding sites (for the same type of sugar residue) per subunit of banana lectin. Molecular cloning revealed that BanLec has sequence similarity to previously described lectins of the family of JRL, and according to molecular modeling studies it has the same overall fold and three-dimensional structure [14,22].

BanLec is reported to be the only one among the lectins of known structure with β-prism I fold to exhibit more than one primary combining site per subunit [22]. It is known that all forms of monocot plants invariably have the β-prism II fold lectins with three carbohydrate-binding sites in each subunit/domain. Whilst β-prism I fold lectins of known structure are all from dicots and they exhibit one carbohydrate-binding site per subunit/domain. However the BanLec which is from a monocot, has β-prism fold I lectin structure with two very similar carbohydrate-binding sites [23].

BanLec furnishes a unique example in proteins where subunit association is not a result of primary hydrophobic interactions. It was reported by Gupta et al. [22], that BanLec is very stable and is denatured only at high concentrations of chaotropic agents. The main reason behind the stability of banana lectin dimer might be attributed to strong hydrogen bonds at the dimeric interface along with the presence of water bridges.

5. BANLEC—PROPERTIES AND BINDING SPECIFICITY

BanLec, the mannose-specific lectin from banana (Musa acuminata, or polyploidy hybrid cultivars thereof), recognizes terminal α-d-mannosyl/glycosyl units, internal α-1,3-mannosyl/glucosyl units, and reducing terminal β-1,3-glucosyl units such as those occuring in laminarin, the polysaccharide from the brown alga Laminaria digitata. This specificity renders it of potential value in medicinal applications, such as inhibition of HIV replication through interaction with its gp120 coat glycoprotein, as has recently been demonstrated [16,24].

The crystallographic, modeling, and simulation studies conducted by Sharma et al. [25], provide fresh insights into the geometrical details of banana lectin–carbohydrate interactions. The lectin associates at the primary binding site with disaccharides involving mannose or glucose residues on α-1,3 linkage towards the non-reducing end. The reducing end occupies the site when the linkage is β-1,3. Both arrangements appear to be allowed to nearly the same extent when the linkage is α-1,2. Interestingly α-1,3-linked disaccharides cannot bind to galactose-specific β-prism I fold lectins, with the non-reducing end at the primary site on account of steric clashes with an aromatic residue that is present only when the lectin is galactose-specific. BanLec is unique in its specificity for internal β-1,3 linkages as well as β-1,3 linkages at the reducing termini [26]. BanLec also explicitly binds to terminal nonreducing α-d-glucosyl/mannosyl units of oligo-/polysaccharide chain ends [21,27,28]. Mo et al. [27], performed extensive experiments in order to find out the exact specificity of BanLec (Musa acuminata) and lectin from the closely related plantain (Musa spp.) by using the techniques of quantitative precipitation, hapten inhibition of precipitation, and isothermal titration calorimetry and showed that they are mannose/glucose binding proteins with a preference for the α-anomeric form of these sugars. Both generate precipitin curves with branched chain α-mannans (yeast mannans) and α-glucans (glycogens, dextrans, and starches), but not with linear α-glucans containing only α-1,4- and α-1,6- glucosidic bonds (isolichenan and pullulan). Another novel observation was also made that banana and plantain lectins recognize their specific binding sites that occur in the linear polysaccharides of elsinan and nigeran, which is unlike that of other mannose/glucose binding lectins (concanavalin A and lectins from pea and lentil).

Again Goldstein et al. [29], further showed that this lectin also binds to the reducing glucosyl groups of β-1,3-linked glucosyl oligosaccharides (e.g., laminaribiose oligomers). Additionally, banana lectin also recognizes β-1,6-linked glucosyl end groups (gentiobiosyl groups) which occur in many fungal β-1,3/1,6-linked polysaccharides. This behavior clearly distinguishes the banana lectin from other mannose/glucose binding lectins, such as concanavalin A and the pea, lentil and Calystegia sepium lectins. Therefore BanLec is unique in its recognition of internal glucosyl/mannosyl residues linked α-1,3-, but not β-1,3-, and especially in its recognition of the reducing sugar unit of laminaribiose (Glcβ1,3Glc) and its higher homologues [27,28,29].

An X-ray analysis of BanLec was undertaken by Singh et al. [18,30], owing to its interesting sugar-binding properties and its similarity to and differences from other JRLs. It was subsequently shown that BanLec is mannose/glucose-specific, with a preference for α-anomeric forms. In addition to α-1-3-linked glucosyl residues, BanLec recognizes reducing glucosyl groups of β-1-3-linked glucosyl oligosaccharides and β-1-6-linked glucosyl end-groups [18,27,29]. Cheung et al. [31], reported another banana lectin from Del Monte bananas (Musa acuminata) having fructose as the most potent inhibitory sugar.

6. BIOLOGICAL PROPERTIES

BanLec is a mannose-specific lectin which readily agglutinates rabbit erythrocytes. BanLec exhibits a strong mitogenic activity towards murine T-cells. The relevance of the mitogenicity of the banana lectin in terms of their physiological role and the impact on food safety was also discussed by Peuman et al. [14], Singh et al. [18] and Sankaranarayanan et al. [19]. BanLec is also reported to act as a potent inhibitor of HIV replication [21,32]. This lectin binds to high mannose carbohydrate structures, including those found on viruses containing glycosylated envelope proteins such as human immunodeficiency virus type-1 (HIV-1). Therefore, Swanson et al. [32], hypothesized that BanLec might inhibit HIV-1 through binding of the glycosylated HIV-1 envelope protein, gp120. They also found that BanLec inhibits primary and laboratory-adapted HIV-1 isolates of different tropisms and subtypes with IC_{50} values in the low nanomolar to picomolar range. The mechanism of BanLec-mediated antiviral activity was explained by their data indicating that BanLec inhibits HIV-1 infection by binding to the glycosylated viral envelope and blocking cellular entry. They also compared the relative favorable anti-HIV activity of BanLec towards other anti-HIV lectins, such as snowdrop lectin and griffithsin, and to anti-HIV drugs currently in clinical use (T-20 and maraviroc). Based on these results, they concluded that BanLec is a potential anti-viral microbicide that could be used to prevent sexual transmission of HIV-1.

BanLec is analogous to receptors present in the plasma and blood cells of crayfish and insect species which recognize and bind to β1,3-glucans [9]. A mannose-specific plant lectins from the Amaryllidaceae family also showed microbicidal property for preventing HIV infection [33]. Balzarini et al. [24], showed that the α-(1-3)-d-mannose- and α-(1-6)-d-mannose-specific agglutinins (lectins) from Galanthus nivalis, Hippeastrum hybrid, Narcissus pseudonarcissus, and Listera ovata inhibited infection of MT-4 cells by human immunodeficiency virus types 1 and 2 (HIV-1 and HIV-2) and simian immunodeficiency virus. They also suggested that these plant lectins interfere with an event in the HIV replicative cycle (fusion process). It explains why BanLec possesses anti-HIV property as it also has specificity towards similar internal α-1-3-linked mannosyl/glucosyl residues.

Table 1. Biological properties of various reported banana lectins.

Name	Source	Specificity	Biological Properties	Authors
BanLec	*Musa acuminata*	Mannose binding	Inhibits HIV-1 infection by binding to the glycosylated viral envelope and blocking cellular entry	Swanson [32]
BanLec	*Musa paradisiaca*	Mannose binding	Stimulates T-cell proliferation	Koshte [12]
BanLec	*Musa paradisiaca*	Mannose binding	Adipogenic effect on mesenchymal cells similar to that mediated by insulin. Induces a mitogen-activated protein kinase (MEK)-dependent ERK signaling in M210B4 cells.	Hinge [35]; Bajaj [36]
BanLec	*Musa basjoo* cv	Glucose/mannose-specific	Macrophage-stimulating, antiproliferative and HIV-1 reverse transcriptase inhibiting activities.	Wong [34]
BanLec	*Musa acuminata*	Mannose/glucose binding protein	T-cell mitogen and induces the formation of IgG4 antibodies.	Mo [27]; Winter [28]
BanLec	*Musa acuminata*	Mannose binding (with two sugar binding sites)	Functions as a potent T-cell mitogen	Meagher [9]
Recombinant BanLec (rBanLec)	Produced in *E. coli*	Similar specificity with native BanLec from *Musa acuminate*	Having potential for modulation of the immune response. Increases T lymphocyte proliferation and secretion of interferon-gamma by mouse splenocytes.	Gavrovic-Jankulovic [26]; Dimitrijevic [37]; Dimitrijevic [38]
BanLec	*Musa acuminata*	Fructose as most potent inhibitory sugar	Mitogenic in murine splenocytes and inducing cytokines interferon-gamma, tumor necrosis factor-alpha, and interleukin-2 expression in splenocytes. Inhibits proliferation of leukemia (L1210) and hepatoma (HepG2) cells and the activity of HIV-1 reverse transcriptase.	Cheung [31]

Cheung et al. [31], reported a lectin from Del Monte banana (Musa acuminata) with cytokine-inducing activity in murine splenocytes. It was able to induce the expression of cytokines such as interferon-gamma, tumor necrosis factor-alpha, and interleukin-2. The lectin also inhibited proliferation of leukemia (L1210) cells and hepatoma (HepG2) cells and the activity of HIV-1 reverse transcriptase. Wong et al. [34], also isolated another banana lectin from Musa basjoo cv. (Emperor Banana) demonstrating stimulatory effects on uptake of [3H-methyl]-thymidine by mouse splenocytes and nitric oxide production by mouse macrophages. The lectin inhibited proliferation of leukemia cells (L1210) and the activity of HIV-1 reverse transcriptase. Hinge et al. [35], and Bajaj et al. [36], showed that with respect to gene expression as well as the signaling mechanisms involved, BanLec from Musa paradisiaca produced an adipogenic action on mesenchymal cells like to that mediated by insulin. They also suggested that this banana lectin may serve as micronutrients for therapeutic purposes in hematological deficiencies. In another paper, Bajaj et al. [36], showed that BanLec interacted with insulin receptors on M210B4 cells and induced a mitogen-activated protein kinase (MEK)-dependent ERK signaling in these cells. Biological properties of various reported BanLec are summarized in Table 1.

7. RECOMBINANT BANLEC

Gavrovic-Jankulovic et al. [26], cloned the gene of banana lectin, and a recombinant protein was produced in Escherichia coli. The cDNA obtained revealed a novel banana lectin isoform, with an open reading frame of 426 nucleotides, encoding a cytoplasmatic protein of 141 amino acids (Table 2). The molecular mass and sugar specificity of the recombinant lectin was comparable with the natural banana lectin. However, Wearne et al. [16], observed that recombinant BanLec (rBanLec) expressed in Escherichia coli, although chemically and functionally identical to native BanLec, differed slightly in its apparent molecular size, probably because rBanLec may be folded differently from the native protein. This rBanLec is a useful reagent for glycoproteomics and lectin microarrays, with a potential for modulation of the immune response (a potent modulator of the proliferation response in CD3+, CD4+, and CD8+ populations of human PBMCs) [16].

Dimitrijevic et al. [37], conducted studies on thermal stability and digestibility of rBanLec on simulated gastro-intestinal fluid as well as simulated intestinal fluid and concluded that rBanLec is a good candidate for the novel bioadhesive lectin-based drug delivery systems to the gastro-intestinal tract (GIT). Dimitrijevic et al. [38], also used rBanLec to study its effects as a mucosal immunostimulator in vivo in mice and found that it is stable in the mouse digestive tract, where it specifically interacts with the mucosal surfaces which can be inhibited by the addition of glucose. They also suggested that rBanLec is a potential oral adjuvant to the gastrointestinal tract.

Table 2. Structural aspects of various reported banana lectins.

Name	Source	Structure	Amino Acid Sequence	Authors
BanLec	*Musa paradisiac*	Homodimeric, ~13-kDa subunits	–	Koshte [12]
BanLec	*Musa acuminata* L.	Homodimeric, 15-kDa subunits	141aa residues	Peumans [14]
BanLec	*Musa basjoo* cv	Homodimeric, 15-kDa subunits	*N*-terminal amino acid sequence similar to lectins from other *Musa* species	Wong [34]
rBanLec	Produced in *E. coli*	cDNA—426 nucleotides 15.9-kDa	141aa residues with *N*-terminal sequences of TGAIKVGA	Gavrovic-Jankulovic [26]; Dimitrijevic [37]

8. DISCUSSION

BanLec are homodimeric and display a molecular weight of about 13–15-kDa (Table 2). Most of them are mannose-specific [9,27,28] except for one species, Musa acuminata, which has fructose [31] as the most potent inhibitory sugar. All of them belong to one family of lectins that is JRLs. BanLec differs from other related members of the JRLs family in term of quaternary structure that involves a unique dimer and is included in the fourth category of JRLs family. Banana lectins are found to possess many exploitable biological properties like HIV-1 reverse transcriptase inhibiting activity, antiproliferative activity toward cancer cells and macrophage stimulating activity (Table 1).

It is well known that bananas are rich in potassium and beneficial due to their health-promoting effects [31]. Because the binding, toxicity, and anti-HIV activity of lectins vary, the identification of novel anti-viral lectins, such as BanLec, will further increase the possibility of successful development of a lectin-based anti-HIV microbicide [32]. rBanLec is also a candidate for the novel oral adjuvant to the gastrointestinal tract. Hence, potential applications could also be explored for rBanLec in view of its safety for human and animal consumption, its resistance to proteolytic degradation and binding to mucosal surfaces, which could increase the availability of therapeutics and enhance or redirect an immune response against target immunogens [26,37,38]. Moreover BanLec is an excellent tool for glycoprotein research [12].

There are various reports on the biomedical applications of BanLec. However, due to the low yield of BanLec from their sources of when conventional methods of isolation are employed, BanLec does not find its glamour in industrial applications. Therefore it is high time to focus more on the development of new methods like genetic engineering for the mass production of rBanLec. Also, more sequence information on BanLec would be helpful for structural studies and functional characterization. BanLec has potential applications based on immunomodulatory, antiproliferative, and antiviral/antimicrobial activities, etc. Nevertheless, there is still a need for other techniques like proteomics and genomic analysis to explore various aspects of these important biomolecules.

ACKNOWLEDGMENTS

We would like to thank DBT, Govt. of India for the support in the publication of the manuscript.

REFERENCES

1. Pachuau, L.; Atom, A.D.; Thangjam, R. Genome classification of Musa cultivars from Northeast India as revealed by ITS and IRAP markers. Appl. Biochem. Biotechnol. **2014**, 172, 3939–3948.
2. Davey, M.W.; Gudimella, R.; Harikrishna, J.A.; Sin, L.W.; Khalid, N.; Keulemans, J. A draft Musa balbisiana genome sequence for molecular genetics in polyploid, inter- and intra-specific Musa hybrids. BMC Genomics **2013**, 14.
3. Azhar, M.; Heslop-Harrison, J.S. Genomes, diversity and resistance gene analogues in Musa species. Cytogenet. GenomeRes. **2008**, 121, 59–66.
4. Uma, S.; Siva, S.A.; Saraswathi, M.S.; Manickavasagam, M.; Durai, P.; Selvarajan, R.; Sathiamoorthy, S. Variation and intraspecific relationships in Indian wild Musa balbisiana (BB), population as evidenced by random amplified polymorphic DNA. Genet. Resour. Crop Evol. **2006**, 53, 349–355.
5. Pachuau, L.; Thangjam, R. Genome characterization of banana genetic resources of Mizoram, India. Sci. Vis. **2012**, 12, 32–36.

6. Sampath Kumar, K.P.; Bhowmik, D.; Duraivel, S.; Umadevi, M. Traditional and medicinal uses of Banana. J. Pharmacogn. Phytochem. **2012**, 1, 57–70.
7. Jain, P.; Bhuiyan, M.H.; Hossain, K.R.; Bachar, S.C. Antibacterial and antioxidant activities of local seeded banana fruits. Afr. J. Pharm. Pharmacol. **2011**, 5, 1398–1403.
8. Tan, E.S.; Aminah, A.; Khalid, H.M.; Mohammad, Y.M.; Maaruf, A.G. Antioxidant properties of three banana cultivars (Musa acuminata "Berangan", "Mas" and "Raja") extracts. Sains Malays. **2012**, 41, 319–324.
9. Meagher, J.L.; Winter, H.C.; Ezell, P.; Goldstein, I.J.; Stuckey, J.A. Crystal structure of banana lectin reveals a novel second sugar binding site. Glycobiology **2005**, 15, 1033–1042.
10. Sharon, N.; Lis, H. Lectins as cell recognition molecules. Science **1989**, 246, 227–234.
11. Raval, S.; Gowda, S.B.; Singh, D.D.; Chandra, N.R. A database analysis of jacalin-like lectins: Sequence-structure-function relationships. Glycobiology **2004**, 14, 1247–1263.
12. Koshte, V.L.; van Dijk, W.; van der Stelt, M.E.; Aalberse, R.C. Isolation and characterization of BanLec-I, a mannoside-binding lectin from Musa paradisiaca (banana). Biochem. J. **1990**, 272, 721–726.
13. Koshte, V.L.; Aalbers, M.; Calkhoven, P.G.; Aalberse, R.C. The potent IgG4-inducing antigen in banana is a mannose-binding lectin, BanLec-I. Int. Arch. Allergy Immunol. **1992**, 97, 17–24.
14. Peumans, W.J.; Zhang, W.; Barre, A.; Houlès Astoul, C.; Balint-Kurti, P.J.; Rovira, P.; Rougé, P.; May, G.D.; van Leuven, F.; Truffa-Bachi, P.; et al. Fruit-specific lectins from banana and plantain. Planta **2000**, 211, 546–554.
15. Clendennen, S.K.; May, G.D. Differential gene expression in ripening banana fruit. Plant Physiol. **1997**, 115, 463–469.
16. Wearne, K.; Winter, H.C.; Goldstein, I.J. Isolation of banana lectin —A practical scale procedure from ripe banana fruit. Prep. Biochem. Biotechnol. **2013**, 43, 285–292.
17. Nakamura-Tsuruta, S.; Kishimoto, Y.; Nishimura, T.; Suda, Y. One-step purification of lectins from banana pulp using sugar-immobilized gold nano-particles. J. Biochem. **2008**, 143, 833–839.

18. Singh, D.D.; Saikrishnan, K.; Kumar, P.; Dauter, Z.; Sekar, K.; Surolia, A.; Vijayan, M. Purification, crystallization and preliminary X-ray structure analysis of the banana lectin from Musa paradisiaca. Acta Crystallogr. D Biol. Crystallogr. **2004**, 60, 2104–2106.
19. Sankaranarayanan, R.; Sekar, K.; Banerjee, R.; Sharma, V.; Surolia, A.; Vijayan, M. A novel mode of carbohydrate recognition in jacalin, a Moraceae plant lectin with a beta-prism fold. Nat. Struct. Biol. **1996**, 3, 596–603.
20. Sharma, A.; Vijayan, M. Quaternary association in β-prism I fold plant lectins: Insights from X-ray crystallography, modelling and molecular dynamics. J. Biosci. **2011**, 36, 793–808. [Google Scholar]
21. Khan, J.M.; Qadeer, A.; Ahmad, E.; Ashraf, R.; Bhushan, B.; Chaturvedi, S.K.; Rabbani, G.; Khan, R.H. Monomeric banana lectin at acidic pH overrules conformational stability of its native dimeric form. PLoS One **2013**, 8, e62428.
22. Gupta, G.; Sinha, S.; Surolia, A. Unfolding energetics and stability of banana lectin. Proteins **2008**, 72, 754–760.
23. Sharma, A.; Chandran, D.; Singh, D.D.; Vijayan, M. Multiplicity of carbohydrate-binding sites in β-prism fold lectins: Occurrence and possible evolutionary implications. J. Biosci. **2007**, 32, 1089–1110.
24. Balzarini, J.; Schols, D.; Neyts, J.; van Damme, E.; Peumans, W.; de Clercq, E. α-(1 –3)- and α-(1 –6)-d-Mannose-specific plant lectins are markedly inhibitory to Human Immunodeficiency Virus and Cytomegalovirus infections in vitro. Antimicrob. Agents Chemother. **1991**, 35, 410–416.
25. Sharma, A.; Vijayan, M. Influence of glycosidic linkage on the nature of carbohydrate binding in β-prism I fold lectins: An X-ray and molecular dynamics investigation on banana lectin-carbohydrate complexes. Glycobiology **2011**, 21, 23–33.
26. Gavrovic-Jankulovic, M.; Poulsen, K.; Brckalo, T.; Bobic, S.; Lindner, B.; Petersen, A. A novel recombinantly produced banana lectin isoform is a valuable tool for glycoproteomics and a potent modulator of the proliferation response in CD^{3+}, CD^{4+}, and CD^{8+} populations of human PBMCs. Int. J. Biochem. Cell Biol. **2008**, 40, 929–941.
27. Mo, H.; Harry, C.W.; van Damme, E.J.M.; Peumans, W.J.; Akira, M.; Irwin, J.G. Carbohydrate binding properties of banana (Musa acuminata) lectin I. Novel recognition of internal α1,3-linked glucosyl residues. Eur. J. Biochem. **2001**, 268, 2609–2615.

28. Winter, H.C.; Oscarson, S.; Slättegård, R.; Tian, M.; Goldstein, I.J. Banana lectin is unique in its recognition of the reducing unit of, 3-O-β-glucosyl/mannosyl disaccharides: A calorimetric study. Glycobiology **2005**, 15, 1043–1050.
29. Goldstein, I.J.; Winter, H.C.; Mo, H.; Misaki, A.; van Damme, E.J.; Peumans, W.J. Carbohydrate binding properties of banana (Musa acuminata) lectin-II. Binding of laminaribiose oligosaccharides and β-glucans containing β1,6-glucosyl end groups. Eur. J. Biochem. **2001**, 268, 2616–2619.
30. Singh, D.D.; Saikrishnan, K.; Kumar, P.; Surolia, A.; Sekar, K.; Vijayan, M. Unusual sugar specificity of banana lectin from Musa paradisiaca and its probable evolutionary origin. Crystallographic and modelling studies. Glycobiology **2005**, 15, 1025–1032.
31. Cheung, A.H.; Wong, J.H.; Ng, T.B. Musa acuminata (Del Montebanana) lectin is a fructose-binding lectin with cytokine-inducing activity. Phytomedicine **2009**, 16, 594–600.
32. Swanson, M.D.; Winter, H.C.; Goldstein, I.J.; Markovitz, D.M. A lectin isolated from bananas is a potent inhibitor of HIV replication. J. Biol. Chem. **2010**, 285, 8646–8655.
33. Balzarini, J.; Hatse, S.; Vermeire, K.; Princen, K.; Aquaro, S.; Perno, C.F.; de Clercq, E.; Egberink, H.; Vanden Mooter, G.; Peumans, W.; et al. Mannose-specific plant lectins from the Amaryllidaceae family qualify as efficient microbicides for prevention of Human Immunodeficiency Virus infection. Antimicrob. Agents Chemother. **2004**, 48, 3858–3870.
34. Wong, J.H.; Ng, T.B. Isolation and characterization of a glucose/mannose-specific lectin with stimulatory effect on nitric oxide production by macrophages from the emperor banana. Int. J. Biochem. Cell Biol. **2006**, 38, 234–243.
35. Hinge, A.; Bajaj, M.; Limaye, L.; Surolia, A.; Kale, V. Oral administration of insulin receptor-interacting lectins leads to an enhancement in the hematopoietic stem and progenitor cell pool of mice. Stem Cells Dev. **2010**, 19, 163–174.
36. Bajaj, M.; Hinge, A.; Limaye, L.S.; Gupta, R.K.; Surolia, A.; Kale, V.P. Mannose-binding dietary lectins induce adipogenic differentiation of the marrow-derived mesenchymal cells via an active insulin-like signaling mechanism. Glycobiology **2011**, 21, 521–529.

37. Dimitrijevic, R.; Jadranin, M.; Burazer, L.; Ostojic, S.; Gavrovic-Jankulovic, M. Evaluation of the thermal stability and digestibility of heterologously produced banana lectin. Food Chem. **2010**, 120, 1113–1118.
38. Dimitrijevic, R.; Stojanovic, M.; Micic, M.; Dimitrijevic, L.; Gavrovic-Jankulovic, M. Recombinant banana lectin as mucosal immunostimulator. J. Funct. Foods **2012**, 4, 636–641.

12

CHAPTER

PROTEINS INVOLVED IN DISTINCT PHASES OF COLD HARDENING PROCESS IN FROST RESISTANT WINTER BARLEY (HORDEUM VULGARE L.) CV LUXOR

Iva Hlaváčková [1,2,*], Pavel Vítámvás [2], Jiří Šantrůček [1], Klára Kosová [2], Sylva Zelenková [3], Ilja Tom Prášil [2], Jaroslava Ovesná [2], Radovan Hynek [1] and Milan Kodíček [1]

[1]*Department of Biochemistry and Microbiology, Institute of Chemical Technology Prague, Technická 5, 166 28 Prague 6, Czech Republic*

[2]*Department of Genetics and Plant Breeding, Crop Research Institute, Drnovská 507/73, 161 06 Prague 6, Czech Republic*

[3]*Department of Plant Experimental Biology, Charles University in Prague, Albertov 6, 128 43 Prague 2, Czech Republic*

ABSTRACT

Winter barley is an economically important cereal crop grown in higher latitudes and altitudes where low temperatures represent an important environmental constraint limiting crop productivity. In this study changes in proteome of leaves and crowns in a frost tolerant winter barley cv. Luxor in relation to short and long term periods of cold followed by a brief frost treatment were studied in order to disclose proteins responsible for the cold hardening process in distinct plant tissues. The mentioned changes have been monitored using two dimensional difference gel electrophoresis (2D-DIGE) with subsequent peptide-mapping protein identification.

Regarding approximately 600–700 distinct protein spots detected on 2D gels, there has been found at least a two-fold change after exposure to low temperatures in about 10% of proteins in leaves and 13% of proteins in crowns. Protein and nitrogen metabolic processes have been influenced by low temperature to a similar extent in both tissues while catabolism, carbohydrate metabolism and proteins involved in stress response have been more affected in crowns than in leaves. The range of changes in protein abundance was generally higher in leaves and chloroplast proteins were frequently affected which suggests a priority to protect photosynthetic apparatus. Overall, our data proved existence of slightly different response strategies to low temperature stress in crowns and leaves, i.e., tissues with different biological role. Moreover, there have been found several proteins with large increase in accumulation, e.g., 33 kDa oxygen evolving protein of photosystem II in leaves and "enhanced disease susceptibility 1" in crowns; these proteins might have potential to indicate an enhanced level of frost tolerance in barley.

Keywords: cold shock; cold and frost acclimation; winter hardiness; hardening; chloroplasts; 2D-DIGE; peptide-mapping

1. INTRODUCTION

Barley can be cultivated under harsher conditions than many other crops, mainly due to its high tolerance to drought and low temperatures [1]. Comprehension of mechanisms by which barley copes with cold and frost is essential for breeding new tolerant cultivars and thus for increasing yields in areas of low production.

Like other plants from the tribe Triticeae, barley exists in spring and winter varieties. For winter cultivars, grown in areas where the temperatures during winter drop below zero, the high frost tolerance is crucial. Moreover, these cultivars require external signal for transition from vegetative phase into reproductive phase; this signal can be vernalization or length of photoperiod [2]. The maximum frost tolerance is reached during the vegetative phase of growth by the cold hardening process (cold acclimation), which takes place predominantly in autumn [3]. The first stage of cold hardening begins at low temperatures above 0 °C, while the second stage requires considerably lower but still non-lethal temperatures [4]. Cold hardening leads to alterations in plasma membrane composition, changes in carbohydrate metabolism and secondary metabolism, photosynthesis, enhanced detoxification and oxido-reduction processes, increased production of osmolytes and synthesis of stress-related proteins

[5]. Among the latter, heat shock proteins (HSP) and late embryogenesis abundant (LEA) proteins represent a major class of protective molecules [6]. Proteomic studies concerning abiotic stresses in plants were recently summarized [7]; at low temperatures, an increase in accumulation of glycolytic enzymes, components of photosynthetic electron transport chain, cp29 RNA-binding protein, reactive oxygen species scavenging enzymes, LEA proteins, chaperones and 70 kDa heat shock protein (HSP70) has been observed repeatedly while accumulation of enzymes involved in saccharide anabolism was decreased. Only few proteomic studies dealt with low temperatures effect on plants belonging to tribe Triticeae; in all of them, wheat was used as a model [8–11].

Our knowledge of cold hardening processes in barley is to date based mainly on transcriptomic studies and studies using one dimensional polyacrylamide gel electrophoresis. Among the best characterized cold induced genes, there belong Wcs120[12] and DHN5[7,13]; proteins encoded by these genes might serve as frost tolerance markers [14,15]. It was shown that majority of cold regulated genes in barley reacted as early as short period of cold (1 day) occurred and only a limited number of barley genes displayed lasting response to prolonged cold treatment [16]. About 67% of genes responsive to cold were found to be chloroplast-dependent cold-regulated genes in leaf tissues. Thus the chloroplasts were suggested to play a major role in the control of barley molecular adaptation to cold [17] and they were proposed as a target for crop improvement [18]. The link between chloroplast activity and the susceptibility of the plant to stress together with retrograde plastid-to-nucleus signaling was demonstrated in Arabidopsis thaliana[19].

In winter cereals, it has been shown that survival of a plant depends on the survival of its crown tissues [20–22]. This part of a plant connects roots and stem and contains meristematic tissue which is responsible for growth restoration in the spring. Recently, different transcription patterns of genes involved in pathways responsible for osmolyte production (sucrose, raffinose, γ-aminobutyric acid), sugar signaling (trehalose metabolism) and secondary metabolism (lignin synthesis) were observed in barley leaves and crowns at low temperatures [23].

In our study, response of barley leaves and crowns to low temperatures (1 day cold shock, 21 days cold acclimation, 1 day frost shock) has been studied by 2D-DIGE coupled with peptide fingerprinting protein identification method using MALDI-TOF mass spectrometry. The aim of this study was to characterize processes leading to enhanced winter hardiness in distinct tissues at proteome level.

2. RESULTS AND DISCUSSION

2.1. Proteomic Analysis

Seedlings of winter barley were exposed to "cold shock" (1 day of 3 °C), "cold acclimation" (21 days of 3 °C) and "frost shock" (1 day of −3 °C after, cold acclimation) and the proteins extracted from leaves and crowns were separated using 2D-DIGE (Figure S1 in Supplemental data). On each gel, approximately 1200 protein spots were distinguished (Figure 1) and with regard to gel to gel variance, it was possible to quantify approximately 700 proteins in crowns and 600 proteins in leaves. The lower number of proteins quantified in leaves could be explained by the presence of highly abundant chloroplast proteins (such as Rubisco large and small chains, see Figure 1b—approximately 60 kDa and 20 kDa) in leaves which decreases the possibility to detect low abundant proteins. This difficulty can be solved by Rubisco depletion; however, each manipulation with samples brings also artificial changes in proteins abundance levels. Therefore, the use of non-photosyntetic tissues, such as crowns in cereals, for gel based proteomic approaches can provide a better view on behavior of low abundant proteins that are not detectable in leaf samples.

In total, 90 proteins in crowns and 63 proteins in leaves changed their abundance at least twice in consequence of cold and frost treatment according to Student's t-test ($p = 0.05$, three biological replicates). By employing the peptide fingerprinting method, 48 proteins in crowns and 39 proteins in leaves were identified (Tables 1 and 2, Table S1 in Supplementary data); among them several proteins with documented relation to cold or freezing stress were found, e.g., (HSP70) and chaperonins [24,25]. In order to facilitate comparison with transcriptome studies, corresponding probe set identifiers on Barley 1 Affymetrix chip [16,17,23] have been found for 8 identified proteins (Tables 1 and 2, Table S1 in Supplementary data). Several proteins (e.g., fructokinase in spot n. 1304, 1311) were identified more than once in the same tissue; these repeatedly observed proteins differed in molecular mass and/or pI observed on gel which is likely due to different isoforms of the same protein, protein modifications or intracellular proteolysis. Some of the differently accumulated proteins were not identified as their abundances on gel were not high enough for subsequent peptide mapping (the peptide mass spectra of appropriate quality were not obtained or the peptide mass spectra did not match to any protein in the database).

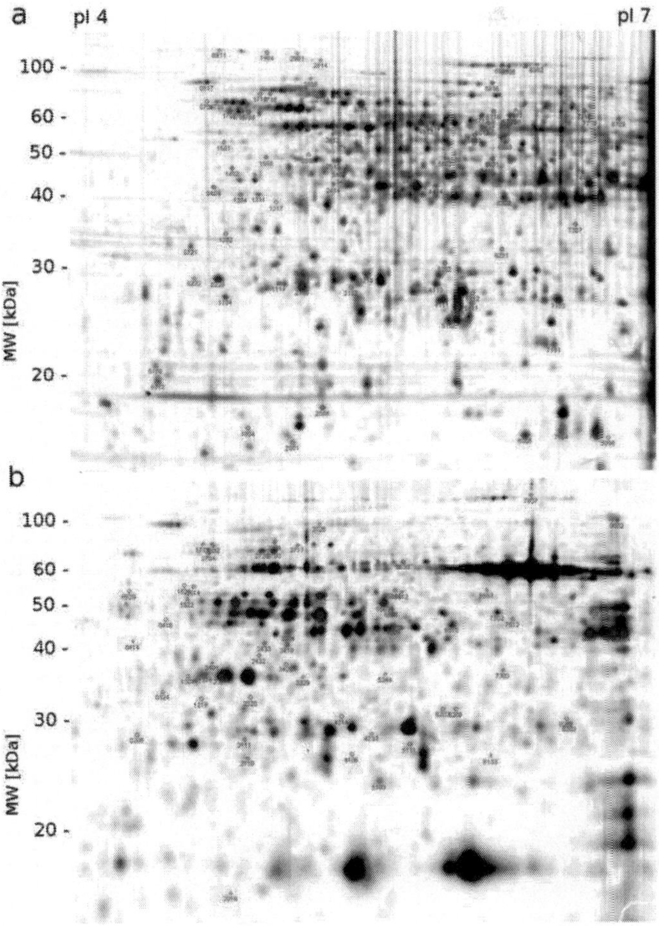

Figure 1. Representative proteome maps of 50 µg total proteins from (**a**) crowns and (**b**) leaves of barley (Hordeum vulgare L.) separated by two dimensional difference gel electrophoresis (2D-DIGE) (24 cm immobilised pH gradient (IPG) strip, pI 4–7, 1 mm thick gel). At least two times up- or down-accumulated proteins ($p = 0.05$, t-test, $n = 3$) are marked with the corresponding spot numbers.

Table 1. List of proteins up- or down- accumulated after low temperature exposure in crowns of winter barley seedlings. Plants were exposed to cold shock, cold acclimation and frost shock; differently accumulated proteins were those which changed abundance at least two times (Student's t-test, p = 0.05, n = 3) in course of low temperature treatment.

Spot No [1]	NCBI identifier (Probe Set ID) [2]	Protein name (organism)	Score [3]	MW_o/pI_o (MW_t/pI_t) [4]	SC (%)/MP [5]	Biological function [6]	Cluster [7]	CL, O [8]
		Crowns						
2814	gi|11527563 (Contig2786_at)	vacuolar proton-ATPase (*Hordeum vulgare*)	187	136/5.4 (69/5.2)	37/20	ATP m. p.	1	
8408	gi|226316443	fructose-bisphosphate aldolase (*Hordeum vulgare*)	120	70/6.7 (39/6.1)	36/11	Carbohydrate m. p.	1	
3404	gi|326490940	malate dehydrogenase (*Hordeum vulgare*)	83	69/5.5 (36/5.6)	25/10	Carbohydrate m. p.	1	
1304	gi|326513418	fructokinase-2 (*Hordeum vulgare*)	139	62/5.2 (36/6.1)	39/17	Carbohydrate m. p.	1	
1311	gi|326513418	fructokinase-2 (*Hordeum vulgare*)	179	62/5.1 (36/6.1)	39/17	Carbohydrate m. p.	1	
4613	gi|357110692	6-phosphogluconate dehydrogenase, decarboxylating-like 1 (*Brachypodium distachyon*)	80	94/5.9 (53/5.6)	22/12	Carbohydrate m. p.	1	
3705	gi|326493636	enolase (*Hordeum vulgare*)	188	107/5.6 (48/5.4)	32/11	Carbohydrate m. p.	1	
2612	gi|326495636	actin (*Hordeum vulgare*)	225	89/5.4 (42/5.3)	47/21	Cytoskeleton organization	1	
108	gi|125548641	h. p. OsI_16233 (*Oryza sativa*)	85	22/4.6 (17/4.8)	47/9	Defence response	1	*
2506	gi|242074406	h. p. SORBIDRAFT_06g029250 (*Sorghum bicolor*)	84	79/5.3 (183/6.0)	11/15	Defence response	1	
12	gi|37651975	chopper chaperone (*Hordeum vulgare*)	75	20/4.6 (11/4.7)	41/6	Metal ion transport	1	
4418	gi|40317416	glutamine synthetase GSr1 (*Triticum aestivum*)	114	69/5.8 (39/5.4)	34/11	Nitrogen compound m. p.	1	
6806	gi|755762 (HVSMEa0004F18f2_s_at)	glutamine synthetase GS2 (*Hordeum vulgare*)	150	137/6.2 (47/5.8)	19/12	Nitrogen compound m. p.	1	c
1713	gi|326496957	thioredoxin peroxidase (*Hordeum vulgare*)	118	113/5.1 (28/6.3)	42/8	Oxidation-reduction p.	1	c
6503	gi|326492375	predicted protein (*Hordeum vulgare*)	77	85/6.2 (41/5.9)	17/7	Oxidation-reduction p.	1	
5611	gi|326493350	isocitrate dehydrogenase (*Hordeum vulgare*)	87	89/6.7 (46/6.0)	20/8	Oxidation-reduction p.	1	
7414	gi|326516152	predicted protein (*Hordeum vulgare*)	122	71/6.6 (39/6.4)	32/14	Oxidation-reduction p.	1	
5716	gi|194707130	unknown (*Zea mays*)	95	104/6.1 (40/8.5)	26/10	Protein catabolic process	1	
2901	gi|326533328	60 kDa chaperonin (*Hordeum vulgare*)	98	141/5.3 (62/5.4)	16/10	Protein folding/refolding	1	c
221	gi|113603 (Contig86_at)	thiol protease aleurain (*Hordeum vulgare*)	76	45/4.8 (40/6.4)	20/7	Protein m. p.	1	
4714	gi|297791831	zinc carboxypeptidase family protein (*Arabidopsis lyrata*)	79	100/5.8 (53/6.9)	22/12	Protein m. p.	1	
2008	gi|326493798	glycine rich protein, RNA binding protein (*Hordeum vulgare*)	139	18/5.4 (17/5.4)	50/10	Response to cold	1	
4304	gi|326494746	predicted protein (*Hordeum vulgare*)	148	63/5.7 (38/5.7)	37/12	Seed storage protein	1	
3601	gi|357149925	elongation factor Tu (*Brachypodium distachyon*)	148	93/5.5 (51/5.9)	40/23	Translational elongation	1	c
5506	gi|326514754	elongation factor Tu (*Hordeum vulgare*)	124	80/6.0 (48/6.0)	48/18	Translational elongation	1	
4614	gi|125580018	h. p. OsJ_36811 (Oryza sativa)	74	95/5.9 (32/6.8)	42/10	Unknown	1	
6303	gi|326491885	predicted protein (*Hordeum vulgare*)	158	61/6.2 (37/6.0)	39/14	Unknown	1	
8303	gi|326497973	predicted protein (*Hordeum vulgare*)	153	61/6.6 (39/6.3)	55/15	Unknown	1	
2808	gi|326505132	predicted protein (*Hordeum vulgare*)	105	123/5.3 (56/4.7)	18/12	Unknown	1	
4414	gi|326517467	predicted protein (*Hordeum vulgare*)	142	65/5.9 (36/5.7)	51/16	Unknown	1	
7717	gi|326517784	predicted protein (*Hordeum vulgare*)	105	104/6.6 (47/6.1)	32/14	Unknown	1	
5416	gi|356532429	LOC100792980 (*Glycine max*)	86	74/6.1 (75/7.2)	18/11	Unknown	1	
5111	gi|400094 (Contig1675_s_at)	JIP-23 (*Hordeum vulgare*)	84	32/6.5 (23/5.9)	38/8	Unknown	1	
2414	gi|61889374 (Contig112_at)	C2 domain-containing protein (*Hordeum vulgare*)	86	66/5.4 (33/5.6)	21/7	Unknown	1	
7012	gi|326493636	enolase (*Hordeum vulgare*)	79	15/6.6 (48/5.4)	32/11	Carbohydrate m. p.	2	
1402	gi|149939605	enhanced disease susceptibility 1 (*Arabidopsis lyrata*)	75	73/5.0 (70/6.1)	15/11	Defence response	2	*
4110	gi|242074406	h. p. SORBIDRAFT_06g029250 (*Sorghum bicolor*)	88	36/5.9 (183/6.0)	9/12	Defence response	2	*
5202	gi|146760217 (Contig1508_s_at)	beta tubulin 6 (*Hordeum vulgare*)	132	41/5.9 (51/4.8)	26/16	Microtubule-based mov.	2	
5521	gi|3915083	beta tubulin (*Cicer arietinum*)	74	78/6.1 (51/4.8)	26/12	Microtubule-based mov.	2	
704	gi|357113738	40S ribosomal protein SA-like (*Brachypodium distachyon*)	185	108/4.9 (33/4.9)	44/16	Proteosynthesis	2	
226	gi|326526175	predicted protein (*Hordeum vulgare*)	82	37/4.9 (26/6.2)	33/7	Response to cold	2	c
		Crowns						
2001	gi|326506132	70 kDa heat shock protein (*Hordeum vulgare*)	257	15/5.2 (74/5.0)	12/8	Response to cold	2	
1705	gi|326492960	70 kDa heat shock protein (*Hordeum vulgare*)	82	102/5.1 (74/5.0)	12/8	Response to cold	2	c,*
2108	gi|326504940	predicted protein (*Hordeum vulgare*)	84	35/5.3 (26/6.2)	33/7	Response to cold	2	
817	gi|326497759	predicted protein (*Hordeum vulgare*)	86	120/4.9 (42/6.0)	28/11	Unknown	2	
5503	gi|326497759	predicted protein (*Hordeum vulgare*)	87	77/5.9 (42/6.0)	20/9	Unknown	2	
2201	gi|326494312	predicted protein (*Hordeum vulgare*)	160	39/5.3 (26/6.2)	33/7	Protein folding	3	
1004	gi|326505132	predicted protein (*Hordeum vulgare*)	175	15/5.4 (39/6.4)	32/14	Unknown	3	*

[1]—spot number on 2D gel; [2]—Probe Set ID on Barley 1 Affymetrix chip; [3]—Mascot score for protein identification (cut off 72); [4]—molecular weight and pI: o—observed on 2D gel and t—theoretical; [5]—Sequence Coverage (%)/Matched Peptides; [6]—biological function according to Gene Ontology annotation; [7]—membership to cluster according to Hierarchical cluster analysis, [8]—Celullar Localisation in chloroplasts—"c" and proteins with excessive change in abundance (outliers, at least 6 fold change)—"*".

Table 2. List of proteins up- or down- accumulated after low temperature exposure in leaves of winter barley seedlings. Plants were exposed to cold shock, cold acclimation and frost shock; differently accumulated proteins were those which changed abundance at least two times (Student's t-test, p = 0.05, n = 3) in course of low temperature treatment.

Spot No [1]	NCBI identifier (Probe Set ID) [2]	Protein name (organism)	Score [3]	MW_o/pI_o (MW_t/pI_t) [4]	SC (%)/MP [5]	Biological function [6]	Cluster [7]	CL, O [8]
			Leaves					
6203	gi|296012105	granule-bound starch synthase I (*Solanum caripense*)	73	90/5.8 (37/6.7)	33/9	Biosynthetic process	1	
3717	gi|32400802	phosphoglycerate mutase (*Triticum aestivum*)	78	114/5.3 (30/5.4)	28/6	Carbohydrate m. p.	1	
5111	gi|303283124	pyruvate kinase (*Micromonas pusilla CCMP1545*)	76	48/5.5 (58/6.0)	21/11	Carbohydrate m. p.	1	
7837	gi|242095346	h. p. SORBIDRAFT_10g009020 (*Sorghum bicolor*)	73	101/6.0 (68/8.8)	14/9	Cell wall biogenesis	1	
5344	gi|157326682	1(10),5-germacradien-4-ol synthase (*Pinus sylvestris*)	76	34/5.7 (73/5.7)	17/11	Metabolic process	1	c
6208	gi|297791885	adenylate kinase (*Arabidopsis lyrata*)	73	90/5.8 (33/9.1)	22/6	Nucleobase-containing compound m. p.	1	c
3504	gi|22795033	putative cytochrome P450 (*Populus tremula*)	96	70/5.3 (24/7.1)	31/7	Oxidation-reduction p.	1	
2327	gi|116787193	unknown (*Picea sitchensis*)	77	43/5.2 (33/9.3)	27/9	Protein m. p.	1	c
3704	gi|75114857	ATP-dependent zinc metalloprotease FTSH 2 (*Oryza sativa Japonica*)	73	89/5.4 (73/5.5)	15/8	Protein m. p.	1	c
3429	gi|225465175	60S ribosomal protein L3-like (*Vitis vinifera*)	82	59/5.4 (25/10.0)	45/9	Proteosynthesis	1	
1307	gi|297609078	Os09g0133600 (*Oryza sativa Japonica Group*)	78	52/5.0 (27/5.1)	28/6	Response to cold	1	
8203	gi|15236846	peroxygenase 1 (*Arabidopsis thaliana*)	82	49/6.3 (28/5.8)	34/12	Response to freezing	1	
2110	gi|212724116	uncharacterised protein (*Zea mays*)	74	116/5.2 (42/9.1)	18/8	tRNA processing	1	
2111	gi|37196631	resistance protein candidate (*Helianthus annuus*)	75	19/5.1 (18/5.1)	31/7	Unknown	1	
1219	gi|296081931	unnamed protein product (*Vitis vinifera*)	74	25/4.9 (242/6.5)	5/11	Unknown	1	
4106	gi|255070529	AAA ATPase (*Micromonas sp. RCC299*)	90	117/5.4 (84/8.0)	21/14	Diverse cellular processes	2	*
5612	gi|226505792	UDP-glucose 6-dehydrogenase (*Zea mays*)	73	85/5.7 (53/6.1)	23/10	Oxidation-reduction p.	2	*
4214	gi|115778	chlorophyll a-b binding protein 1 (*Sinapis alba*)	73	128/5.5 (24/5.2)	27/7	Photosynthesis	2	c
2401	gi|115436780	33kDa oxygen evolving protein of photosystem II (*Oryza sativa Japonica Group*)	79	53/5.2 (35/6.1)	23/5	Photosynthesis	2	c,*
1521	gi|242035457	Sb01g032420 (*Sorghum bicolor*)	74	59/5.2 (80/5.6)	12/10	Transport	2	
6133	gi|145351306	F-ATPase (*Ostreococcus lucimarinus*)	75	76/5.8 (68/5.0)	17/9	ATP m. p.	3	
			Leaves					
8202	gi|29893610	SR-rich pre-mRNA splicing activator (*Oryza sativa*)	75	143/6.0 (23/5.9)	36/7	mRNA processing	3	
529	gi|255560267	chaperonin 60 kDa (*Ricinus communis*)	96	91/4.7 (61/6.2)	25/14	Protein folding/refolding	3	
2320	gi|15236846	peroxygenase 1 (*Arabidopsis thaliana*)	83	39/5.2 (28/5.8)	30/8	Response to freezing	3	
3329	gi|255070529	AAA ATPase (*Micromonas sp. RCC299*)	74	112/5.6 (84/8.0)	12/8	Diverse cellular processes	4	*
8414	gi|255555035	ATP binding protein (*Ricinus communis*)	74	81/6.2 (75/5.8)	21/10	MAPK cascade	4	
7502	gi|297607677	kinesin (*Oryza sativa*)	75	76/6.0 (41/5.4)	33/14	Microtubule-based mov.	4	
2510	gi|755762 **(HVSMEa0004F18f2_s_at)**	glutamine synthetase GS2 (*Hordeum vulgare*)	101	66/5.2 (47/5.8)	46/10	Nitrogen compound m. p.	4	c
			Leaves					
4233	gi|48093455	ADH-like UDP-glucose dehydrogenase (*Nicotiana tabacum*)	75	36/5.5 (42/6.2)	24/10	Oxidation-reduction p.	4	
205	gi|115793 **(Contig636_at)**	chlorophyll a-b binding protein of LHCII type III (*Hordeum vulgare*)	73	43/4.8 (29/5.0)	26/6	Photosynthesis	4	c
3410	gi|297727991	Os11g0153600 (*Oryza sativa*)	73	113/5.2 (27/10.8)	41/7	GTP catabolic process	4	c
3622	gi|226508704	elongation factor Tu (*Zea mays*)	101	75/5.3 (51/6.1)	33/15	Translational elongation	4	c
5611	gi|23476281	myb-like transcription factor 1 (*Gossypium raimondii*)	75	59/5.7 (31/7.1)	19/6	Transription	4	
2432	gi|255635003	unknown (*Glycine max*)	73	60/5.1 (34/6.4)	21/6	Unknown	4	c
5503	gi|225440886	h. p. (*Vitis vinifera*)	74	42/5.8 (18/6.0)	49/8	Unknown	4	
8537	gi|255585735	calmodulin binding protein (*Ricinus communis*)	75	78/6.2 (54/8.9)	26/13	Unknown	4	
9404	gi|225464180	h. p. (*Vitis vinifera*)	74	46/6.5 (37/6.1)	25/7	Unknown	4	
2019	gi|147785747	h. p. VITISV_033291 (*Vitis vinifera*)	81	112/4.9 (25/5.3)	38/8	Protein dephosphorylation	5	*
5630	gi|302805059	h. p. SELMODRAFT_423400 (*Selaginella moellendorffii*)	73	83/5.7 (48/8.7)	29/12	Unknown	5	*

[1]—spot number on 2D gel; [2]—Probe Set ID on Barley 1 Affymetrix chip; [3]—Mascot score for protein identification (cut off 72); [4]—molecular weight and pI: o—observed on 2D gel and t—theoretical; [5]—Sequence Coverage (%)/Matched Peptides; [6]—biological function according to Gene Ontology annotation; [7]—membership to cluster according to Hierarchical cluster analysis, [8]—Celullar Localisation in chloroplasts—"c" and proteins with excessive change in abundance (outliers, at least 8 fold change)—"*".

Only one protein was identified in samples from both tissues—glutamine synthetase (gi|755762) with almost similar decrease of abundance after cold in both tissues (spot n. 6806 in crowns and 2510 in leaves); moreover, the glutamine synthetase GSr1 (gi|40317416) isoform in crowns also decreased its abundance.

In total, five members of different protein families have been identified in both tissues; AAA ATPase (gi|255070529 in leaves and gi|194707130 in crowns), ribosomal proteins (60S in leaves—gi|225465175; 40S in crowns—gi|357113738), elongation factor Tu (gi|226508704 in leaves, gi|326514754 and gi|357149925 in crowns), vacuolar proton ATPase (gi|145351306 in leaves, gi|11527563 in crowns) and chaperonin 60 kDa (gi|255560267 in leaves, gi|326533328 in crowns). The noticeable difference between leaf and crown tissue was observed in protein spot 4106 identified as AAA ATPase on leaf gels. Spot 4106 increased its abundance after cold shock and decreased its abundance after cold acclimation in leaves whereas in crowns the abundance of AAA ATPase was constantly decreasing in course of low temperature exposure. The difference was also found in chaperonin 60 kDa; there was an increase in its abundance after cold acclimation in leaves (gi|255560267) whereas in crowns, its abundance was decreased (gi|326533328) at the same time. Different changes in protein abundance of AAA ATPase and chaperonin 60 kDa could be explained by identification of different isoforms with probably slightly different activity under cold stress. However, most of the proteins identified in both tissues showed similar changes in abundance under the cold treatment and this result thus may indicate to some extent analogous regulation of common cold-responsive pathway in cold stressed plant tissues (such as increased dehydrin levels in cold-treated leaves and crowns; [26]).

2.2. Biological Processes Influenced by Low Temperatures in Leaves and Crowns

Biological functions of differently accumulated proteins were determined by their GO annotation in domain of biological processes according to UniProtKB (Tables 1 and 2, Table S1 in Supplementary data). In total, 19 proteins (22% of identified proteins) were assigned to category "Unknown" as their functions have not been determined yet. However, this result indicates that these yet "unknown" proteins are cold-regulated (14 proteins revealed a decreased accumulation and 5 proteins revealed an increased accumulation during cold treatment). The global overview of biological processes affected by low temperatures was obtained using WEGO application; the functional classification using WEGO reflects the fact that a single protein can be involved in various processes and therefore classified in more than one functional category. After setting of GO annotation level to five, 16 functional categories were chosen to cover our data and to pointed out existing differences between leaves and crowns in response to low temperatures (Figure 2, Tables S1 and S2 in Supplementary data).

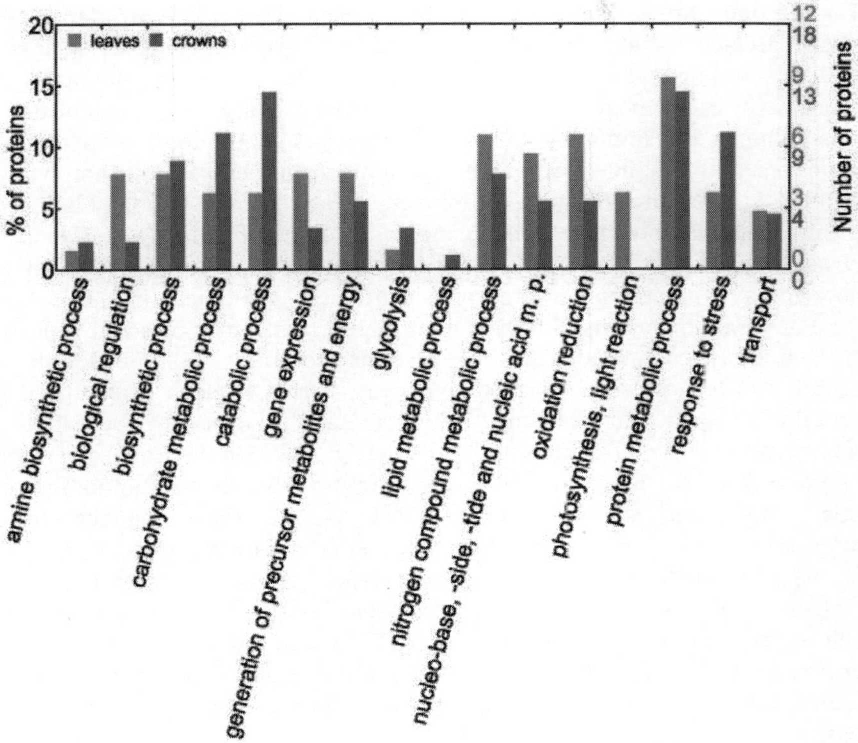

Figure 2. Functional classification of differently accumulated proteins showing a comparison between crowns and leaves of winter barley exposed to cold and frost. In total, 48 proteins in crowns and 39 proteins in leaves were assigned to one or more of 16 categories selected for classification using Web Gene Ontology Annotation Plotting (WEGO) application. On the y axis the percentage of differently accumulated proteins possesing similar biological function in distinct tissues can be seen, on the x axis respective biological processes are listed. WEGO classification reflects the fact that single protein can be involved in several processes, thus sum of % in all categories is higher than 100.

The "protein metabolic processes" represents the largest functional category being formed by approximately 15% of identified proteins both in leaves and crowns. Regulation of protein metabolism; represented by proteases, elongation factors and ribosomal proteins, appears to be of the same significance in leaves and crowns in the response to low temperatures. Number of proteins in this category suggests importance of unfunctional proteins removal and necessity of proteosynthesis regulation.

The second largest category "catabolic processes" (6% of identified differently accumulated proteins in leaves and 14% in crowns) confirmed previous finding that the cold hardening is an active and energy demanding process [27]. The next large category, "nitrogen compound metabolism", was influenced to similar extent in both tissues (11% of identified proteins in leaves and 8% in crowns, e.g., glutamine synthetase) thus implying universal importance of these proteins in stress response to cold. Changes in nitrogen metabolism related proteins accumulation have been so far observed mainly as a consequence of drought and salinity stress [7]. Other noteworthy functional categories contained proteins involved in "carbohydrate metabolism" (6% of identified proteins in leaves and 11% in crowns, e.g., pyruvate kinase) and "response to stress" (6% of identified proteins in leaves and 11% in crowns, e.g., peroxygenase). Interestingly, protein grouped into category "lipid metabolic process" (i.e., lipase "enhanced disease susceptibility 1"—EDS1; gi|149939605, Table S1) was found only in crowns. In this case, it is more probable that accumulation of this protein has also been changed in leaves but these changes were undetectable due to presence of highly abundant photosynthetic proteins in leaves (see subchapter 2.1.). It is not surprising that proteins (33 kDa oxygen evolving protein of photosystem II (PsbO), gi|115436780; chlorophyll a-b binding proteins, gi|115778 and gi|115793; Table S1) grouped into "photosynthesis, light reaction" functional category have been found only in leaves and not in crowns which are non-photosynthetic tissue.

On account of the differences found in numbers and quality of proteins responsible for separate biological processes influenced by low temperatures in distinct tissues, it is reasonable to believe in existence of slightly different response strategies to low temperature stress in crowns (plant survival and growth renewal) and leaves (energy metabolism), i.e., tissues with different biological role [20].

2.3. Proteins with Large Changes in Abundance

Exploratory data analysis revealed outlying proteins in our data sets, the proteins which changed their abundance excessively compared to the rest of differently accumulated proteins. On the basis of the large changes in accumulation of these proteins, we can conclude that they may play substantial roles in frost tolerant winter barley response to low temperature. Since the data sets observed in leaves and crowns differed in their characteristics, the extent of changes in proteins abundances in leaves were greater compared to crowns (Figure S2 and Table S3 in Supplementary data), the limit for outliers (the proteins which displayed excessive changes in abundance) has been set as different values for leaves and crowns, respectively. The boxplot method (2.698 σ) suggested 8 outliers in leaves which changed their abundance at least 8 times and 7 outliers in crowns which changed abundance at least 6 times (Tables 1 and 2, in more detail Tables S1 and S3 in Supplementary data).

In leaves, the outliers were found in all clusters (see subchapter 2.4., Figure 3, Table 2) except Cluster 1 (increase of abundance) and only 5 of the 8 identified proteins have known function (Table 2, Table S1 in Supplementary data). Among them, there was AAA ATPase (gi|255070529) which was found in two spots; in spot 4106, its abundance was 10 times increased after frost shock/cold acclimation while in spot 3329, the abundance was 9 times decreased (Table S1 in Supplementary data). According to position on gel, these two spots differed slightly in molecular weight and pI (spot 4106: 117 kDa, pI 5.37 and spot 3329: 112 kDa, pI 5.6), these differences might be caused by protein modifications in reaction to frost stress. Proteins of AAA family are essential for many cellular functions including degradation of misfolded proteins; previously they have been observed as up-accumulated during cold acclimation [28]. Another highly accumulated protein in leaves was protein PsbO (gi|115436780, 19 times increase after frost shock/cold shock), a part of the oxygen-evolving complex which stabilises manganese. Together with the gene for chloroplast protein PsbO several other genes are co-expressed, including ATPases [29], which is in accordance with our finding. The enzyme UDP-glucose 6-dehydrogenase (gi|226505792), which is involved in diverting UDP-Glc to cell wall biosynthesis among other processes, increased abundance twice after cold shock and then excessively decreased abundance (10 times decreased after frost shock/cold shock). Low temperature stress leads to cell dehydration and causes changes in cell volume and cell compartment organization; therefore, the adjustment of cell wall is a necessity for holding cell integrity; several enzymes participating in UDP-Glc formation and thus cell wall biogenesis have been already observed as up-accumulated after short cold treatments, e.g., UDP-glucose pyrophosphorylase or sucrose synthase 1 [30]. In leaves, the hypothetical protein VITISV_033291" (gi|147785747) was 11 times decreased after cold shock and then 3 times increased both after cold acclimation and frost shock. This protein belongs to protein phosphatase family C (PP2C), one of the largest families in the plant kingdom, and proteins from this family have been found as regulators of signal transduction pathways and also involved in development [31]. It has been shown by antisense approaches that down-regulation of PP2C accelerated plant development and increased freezing tolerance [32]; we observed a similar trend at protein level in our experiment.

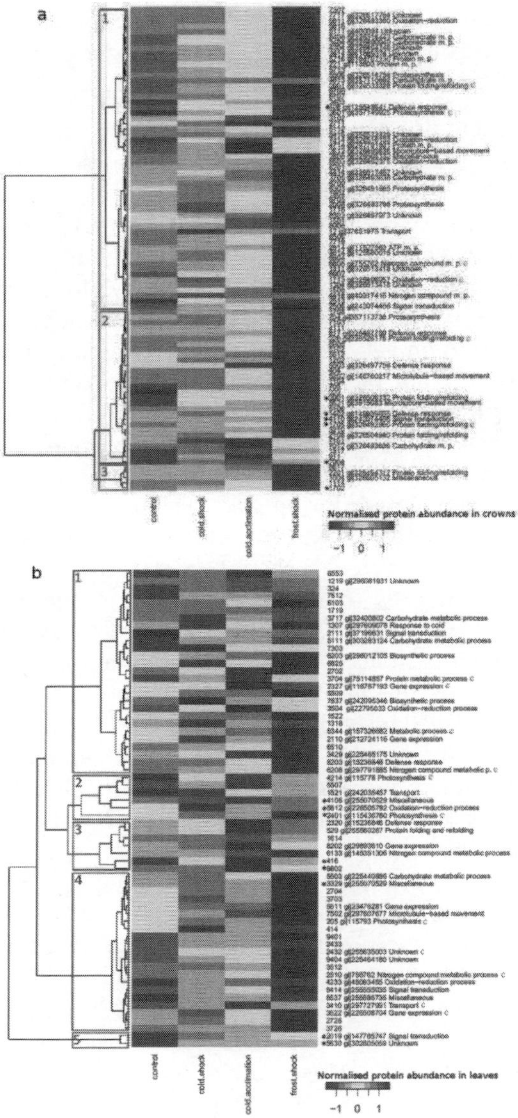

Figure 3. Hierarchical cluster analysis of proteins differently accumulated in (**a**) crowns and (**b**) in leaves in course of low temperature exposure. Red boxes mark clusters of proteins which have similar mode of accumulation; dendrogram was constructed using Ward's method and cut off at height h = 10. The sign * marks proteins which changed their abundance at least 8 times in leaves and 6 times in crowns, letter "c" marks proteins from chloroplasts.

Crown outliers were found in all clusters but most of them belonged to Cluster 2; 5 out of 7 outliers were identified. Amongst the identified proteins with known function, there has been found chloroplast HSP70 (gi|326492960, 7 times increased after frost shock) which is a molecular chaperone involved in protein aggregation prevention and refolding under normal and stress conditions. It also takes part in protein import, translocation and degradation [33]. Another protein with outlying changes in abundance in crowns was EDS1 (gi|149939605, 7 times increased after frost shock compared to cold shock) which controls, among other processes, cell death propagation in response to chloroplast-derived reactive oxygen species [34]. The next two outliers identified in crowns (hypothetical protein OsI_16233 and hypothetical protein SORBIDRAFT_06g029250) are according to GO annotation involved in defense response; the last identified protein (predicted protein gi|326505132) has not been assigned to any function yet.

Interestingly, the extent of changes in protein abundance and the occurrence of outliers were generally higher in leaves, implying a need of an increased level of proteosynthesis and protein degradation during cold acclimation. The fact that the numbers of all differently accumulated proteins were higher in crowns (Table 1) corresponded to the importance of crowns for plant survival [20]. Moreover, outliers found in leaves showed large increase (e.g., 33 kDa oxygen evolving protein of photosystem II) or decrease (e.g., UDP-glucose 6-dehydrogenase) in abundance mainly after cold acclimation and frost shock and they belonged mostly to Cluster 2 and 3 whereas in crowns most of outliers mounted their abundance after frost shock (e.g., EDS1) and they were sorted to Cluster 3. This implies that proteins with large changes in abundance were highly accumulated under cold treatment in crowns (e.g., HSP70, EDS1) contrary to high decrease of some proteins accumulation in leaves (e.g., PsbO, UDP-glucose 6-dehydrogenase, AAA ATPase) in at least one sampling time during cold treatment. From our previous studies on dehydrins, e.g., [7], it is known that crop winter cultivars accumulate some protective compounds such as dehydrins also under optimal temperature of 20 °C and these proteins are accumulated in higher levels in crowns than in leaves [26]. On the other hand, leaves as photosynthetic organs play an important role in an energy-demanding process of cold-acclimation which is needed for plant survival at low temperatures. Therefore, our results indicated that after sensing cold the leaf cells reorganized some of their anabolic and catabolic processes to a higher extent when compared to crown cells; these large changes might be in order to protect photosynthetic apparatus.

The large changes in protein abundance after cold treatment observed at outliers could also indicate a potential of these proteins to evaluate cereal cold tolerance (such as dehydrins, e.g., [35]). However, these cold-regulated proteins should be first tested in a large set of differently cold-tolerant barley genotypes to verify them as possible protein markers.

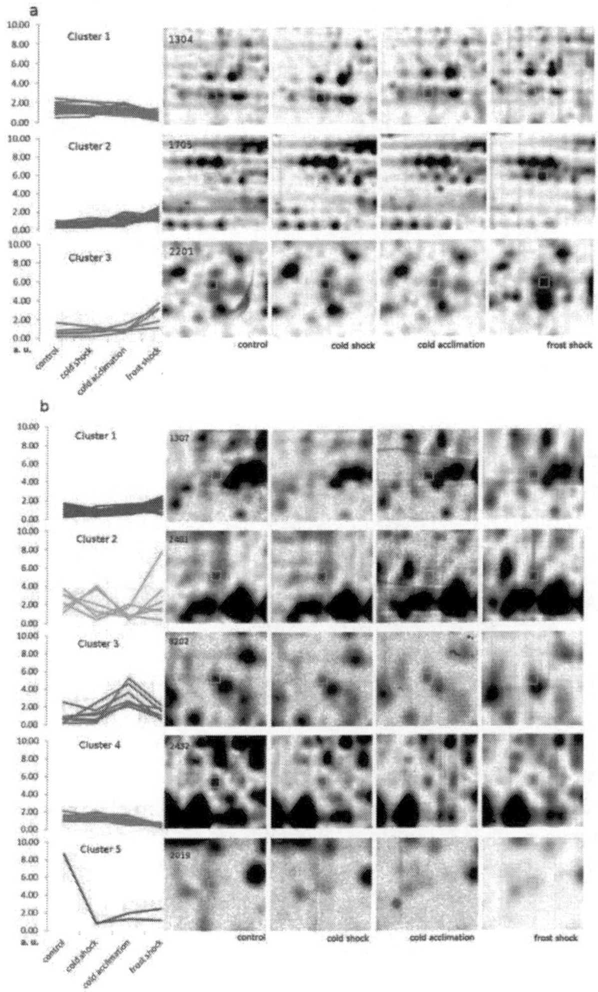

Figure 4. Modes of protein accumulation in (**a**) crowns and (**b**) leaves after low temperature exposure. Vertical axis represents protein abundance and horizontal axis represents different treatments, all proteins in the same cluster are drawn in the same colour. Representative protein spots pictures on 2D-gels for different treatments are given for each cluster; spot number is in left upper corner of the image and corresponding proteins are listed in Tables 1 and 2.

2.4. Protein Accumulation Modes Induced by Low Temperature

Hierarchical cluster analysis sorted differently accumulated proteins to three clusters in crowns and five clusters in leaves (Tables 1 and 2, Figures 3 and 4, Table S1 in Supplementary data). Proteins of which abundances were predominantly increasing in course of low temperature exposure were found in Cluster 1 in leaves (27 proteins/15 identified) and Cluster 2 in crowns (29 proteins/12 identified), the continuously mounting abundance suggests their importance in all phases of cold hardening process. Among these proteins, there are, e.g., HSP70 in crowns and ATP-dependent zinc metalloprotease FTSH 2 in leaves with documented relation to cold stress [6,28].

In Cluster 4 in leaves (21 proteins/13 identified) and in Cluster 1 in crowns (56 proteins/34 identified), there are proteins of which abundances decreased gradually after low temperature exposure, e.g., 23 kDa jasmonate-induced protein (JIP-23) in crowns (gi|400094, Contig1675_s_at); JIP-23 might regulate cell wall polysaccharide synthesis [36]. In barley leaves, JIP-23 genes were expressed upon various stresses such as drought, sorbitol treatment and cold [16,37]. However, transcript (Contig1675_s_at) of the gene belonging to this protein displayed downregulation after prolonged cold treatment (4 °C, 49 day) in barley [16]. Proteins of which abundances have decreased upon cold both in leaves and in crowns include elongation factor Tu and glutamine synthetase (GS1 and GS2). Elongation factor Tu plays a key role in the elongation phase of protein synthesis in plant organelles including mitochondria and plastids and has an important role in stress responses; it has been reported in a transcriptomic study [38] as up-accumulated after cold stress 5/2 °C and at proteomic level as up-accumulated after cold stress 15 °C but no up-accumulation was observed after cold stress of 5 °C [30], which is in agreement with our results. Glutamine synthetase is involved in primary nitrogen assimilation and glutamine generation (isoenzyme form GS1), as well as in NH_4^+ reassimilation released via photorespiration (isoenzyme form GS2) [39]. Glutamine synthetase has been observed to be up-accumulated during salt stress [40,41]; nevertheless, from overexpression experiment [42] it can be concluded that lower abundance of GS1 might have a positive influence on frost resistance. Furthermore, it can be seen that in crowns the key proteins of nitrogen metabolism (glutamine synthetases GS1 and GS2) and almost all proteins related to carbohydrate metabolism (glycolysis and pentose-phosphate shunt) are placed in the Cluster 1, thus they are gradually downregulated in a similar manner. The exception is spot 7012—enolase, which mounted its abundance after cold treatment, but according to its molecular weight observed on gel (Table S1 in Supplementary data), it is most probably only a fragment after degradation. Interestingly, C2 domain-containing protein (Cluster 1; gi|61889374, gradually decreased after cold) has been observed on transcriptomic level (Contig112_at) as upregulated after cold shock in barley seedlings (4 °C, 1 day; [16]). This discrepancy is an illustration of the problematic comparison of transcriptomic results and proteomic results. Generally, the changes in abundance of stress-regulated transcripts between control and stress-treated samples are higher than similar changes observed at protein level (e.g., JIP-23 in [16]). The accumulation of transcript in tissue does not necessarily indicate real abundance of the corresponding protein since there are several levels of translational regulation, the translated proteins can be maintained for a long time (such as dehydrins under cold, e.g., [13,26] or degraded in a short time. Therefore, the changes in gene expression at transcript level do not often correspond with the changes at protein level [7].

Cluster 5 in leaves (2 proteins/2 identified) contains proteins of which abundances were strongly decreased after cold shock and then slightly increased again after cold acclimation; one of the identified proteins was "hypothetical protein VITISV_033291" (gi|147785747) which showed large decrease in its abundance.

Conspicuous changes in abundance induced by either cold shock or frost shock were exhibited mainly by proteins in Cluster 2 in leaves (6 proteins/5 identified), these proteins therefore seem to be of importance during sudden short low temperature stress, e.g., AAA ATPase and chlorophyll a-b binding protein 1 observed previously in [28,43].

Proteins of which abundances increased exclusively after cold acclimation were found in Cluster 3 in leaves (7 proteins/4 identified), e.g., chaperonin 60 kDa, peroxygenase 1. Peroxygenases are enzymes involved in the biosynthesis of phytooxylipins, which play important roles in plant defense responses [44]. Similarly, Cluster 3 in crowns (5 proteins/2 identified) consists of proteins which did not change their abundance after cold shock noticeably but their abundance increased significantly after cold acclimation or frost shock, thus they are likely of considerable role in later phases of cold hardening process. Two identified proteins were "predicted protein" gi|326494312 which is a chaperon responsible for protein folding and "predicted protein" gi|326505132 from protein family von Willebrand factor, type A with RNA-binding activity. Recently, protein LGD1 which possesses von Willebrand factor type A (vWA) domain has been studied in rice. Isoforms of this protein were detected mainly in nucleus and partly in cytoplasm and it was demonstrated that LGD1 regulates rice vegetative growth and development [45].

2.5. Importance of Chloroplasts in the Cold Hardening Process

Cellular localization of identified differently accumulated proteins was determined by Gene Ontology annotation and by TargetP 1.1 server; 6 proteins from crowns and 11 proteins from leaves were localized to chloroplasts (Tables 1 and 2, Tables S1 and S4 in Supplementary data). In leaves, the frequent occurrence of differently accumulated chloroplast proteins was anticipated as they constitute the majority of the total proteins in leaves. To sustain photosynthetic reaction seems to be an important prerequisite for energy dependent cold acclimation process [7]. Both chlorophyll a-b binding proteins (gi|115793 and gi|115778; part of light-harvesting complex) decreased abundance under cold treatment. Under changing conditions, the reversible phosphorylation of chlorophyll a-b binding proteins (LHCII) represents a system for balancing the excitation energy between the photosystems I and II [46]. Therefore, the decrease of spots identified as chlorophyll a-b binding proteins seems to be a result of the reversible phosphorylation following lower level of photosynthesis in cold-treated plants. However, crowns are not photosynthetically active and they are formed by meristematic tissue where only proplastids are present; therefore, a relatively high number of proteins localised to chloroplasts in crown tissue is remarkable. Most probably, these proteins originate from proplastids. A noteworthy fact is that PsbO found in leaves and HSP70 found in crowns, both of chloroplast localisation, have revealed an excessive change in abundance which confirms an important role of chloroplast protein accumulation in the winter barley cold hardening process [17,18].

3. EXPERIMENTAL SECTION

3.1. Plant Samples

Winter barley (Hordeum vulgare L., cultivar Luxor) plants were cultivated as described in Jánská et al.[23]. In short, seedlings were grown in soil under a 12 h photoperiod and an irradiation intensity ~200 μmol $m^{-2}s^{-1}$ at a day/night temperature of 18/13 °C in a growth cabinet. In the second leaf stage, the plants were exposed for up to 3 weeks to a day/night temperature +3/2 °C and irradiation intensity was decreased to ~120 μmol $m^{-2}s^{-1}$. Following the 21 day acclimation period, a set of seedlings was also exposed to frost shock of −3 °C for 1 day. For 2D-DIGE analysis, roughly ten plants for each of all three biological replicates were used to make one sample of crown or a middle part of the second leaf tissue. Thus for each three biological replicates, at least two technical replicates (20 plant) were taken at each sampling time point (0 day—"control" samples, 1 day—"cold shock" samples, 21 day—"cold acclimation" samples and 21 day with 1 day of freezing—"frost shock" samples) from different pots in each growth cabinet.

3.2. Protein Extraction

Total soluble proteins were extracted as described in [47] with some modifications. Approximately 200 mg of plant material were grinded to fine powder in liquid nitrogen and then 1.5 mL of 10% TCA in acetone was added. After one hour at −20 °C, sample was centrifuged 10 min at 12,000× g. Resulting pellet was washed three times with cold 0.07% DTT in acetone and air-dried. Dry pellet was re-suspended in 0.5 mL of SDS buffer (30% sucrose, 2% SDS, 5% 2-mercaptoethanol, 0.1 M Tris-HCl, pH 8.0) and 0.5 mL of phenol (0.5 M Tris-HCl saturated, pH 8.0). Proteins were extracted one hour at room temperature and then centrifuged 10 min at 12,000× g. Upper phenol phase was placed into new microtube and proteins were precipitated with 5 volumes of cold 0.1 M ammonium acetate in methanol overnight at −20 °C. The sample was then centrifuged for 10 min at 12,000× g and washed three times with cold acetone, resulting pellet was air-dried. Proteins were dissolved in lysis buffer (7 M urea, 2 M thiourea, 4% w/v CHAPS, 30 mM Tris, pH 8.0) and the pH of the lysate was adjusted to 8.5 by adding 50 mM sodium hydroxide. Protein concentration was determined using 2D Quant Kit (GE Healthcare Bio-Sciences, Piscataway, NJ, USA).

3.3. 2D-DIGE and Image Analysis

Protein samples were labeled with Amersham CyeDye™ DIGE Fluors (GE Healthcare, Buckinghamshire, UK) according to manufacturer instructions. Covalent binding of dyes to proteins allows separation and subsequent quantitation of up to three samples in one gel. In our experiments, two samples and an internal standard (30 μg of proteins per each sample) were labeled, then mixed together and applied on a strip with an immobilized pH gradient (IPG strip, ReadyStrip™ IPG Strips, 24 cm, pH 4–7, Bio-Rad, Hercules, CA, USA) according to manufacturer instruction. Then isoelectric focusing was carried out on Protean IEF Cell (Bio-Rad, Hercules, CA, USA) with maximal voltage 10,000 V until 75,000 V-hours was reached. After isoelectric focusing, proteins in IPG strips were reduced for 12 min by 1% DTT and alkylated for 12 min by 2.5% IAA dissolved in equilibration buffer (6 M urea, 2% SDS, 20% glycerol, 0.375 M Tris-HCl, pH 8.8). Then, IPG strips were placed on the top of 12.5% polyacrylamide gels (1 mm thick) and sealed with 0.5% low melting agarose. Polyacrylamide gel electrophoresis was performed using Ettan Daltsix Electrophoresis System (GE Healthcare) for approximately 6 hours. Gels with separated proteins labeled with CyeDyes were scanned on Pharos FX Plus (Bio-Rad) at a resolution of 100 μm.

Differently accumulated proteins, i.e., those that changed their abundance at least twice according to Student's t-test ($p = 0.05$, three repetitions), were found using PDQuest 8.0.1 (Bio-Rad) program. The change in protein abundance in a single tissue (in leaves or in crowns) was computed as a ratio of average protein abundance at two distinct sampling dates (e.g., average abundance in leaf sample "cold shock" to average abundance in leaf "control" sample). Only the protein spots that could be detected on three gel replicates were considered for subsequent analysis.

3.4. Protein Identification

Spots containing differently accumulated proteins were cut out from a preparative 2D gel containing 1 mg of protein mixture and stained with Coomassie Brilliant Blue (Bio-Rad). Pieces of gel were de-stained using 100 mM NH_4HCO_3/ACN 1:1 (v/v) and proteins in the gel were digested for 3 h at 37 °C in 50 mM NH_4HCO_3 with trypsin (concentration 12.5 μg/mL, sequencing grade, Promega, Madison, WI, USA). The resulting peptides were extracted with 0.1% TFA and then desalted and concentrated using ZipTip C-18 (Millipore, Billerica, MA, USA). Peptide samples as well as external calibration standard (Bruker peptide calibration standard I, Bruker, Bremen, Germany) were mixed 1:1 with 2,5-dihydroxybenzoic acid (20 mg/mL in 33% ACN and 0.1% TFA; Sigma-Aldrich, Schnelldorf, Germany) and applied on a MALDI target plate. All mass spectrometric analysis used Biflex IV MALDI-ToF mass spectrometer (Bruker, Bremen, Germany) in reflective positive ion mode in the range of 800–4000 Da (300 laser shots for each spectrum). Spectra were analyzed in mMass 3.10.0 program (Praha, Czech Republic) [48]; signal to noise threshold was set to 3. For protein identification, the extracted peaklists were searched using Mascot server against NCBInr database (version 20120623; 18713758 sequences; 6412106995 residues) with the following settings: organism Viridiplantae—green plants; enzyme trypsin (1 misscleavage allowed), cysteine carbamidomethylation was set as a fixed modification and methionine oxidation as a variable modification; peptide tolerance was set to 0.2 Da. This settings cut off score was 72. Since the barley genome is not fully sequenced, a portion of proteins were identified on the basis of homology with proteins from different plant species. Corresponding gene identifiers on the "GeneChip Barley Genome Array" Affymetrix chip were found in the file "Barley1.na32.annot.csv" at Affymetrix's websites [49].

3.5. Biological Functions of Identified Proteins

Biological functions of proteins and their cell localisation were searched in UniProtKB database [50] according to their Gene Ontology annotation (GO) [51]. To confirm cellular localisation of identified proteins with GO annotation and to determine localisation of those without GO annotation, TargetP 1.1 server [52] was used; specificity was set to >0.95.

In order to get a global overview of biological processes influenced by low temperatures, proteins were classified to categories using Web Gene Ontology Annotation Plot (WEGO) [53,54].

3.6. Statistical Analysis of Differently Accumulated Proteins

The differently accumulated proteins were analyzed in a R 2.15.0 environment for statistical computing [55]. In order to summarize main characteristics of data, exploratory data analysis was done using Histogram, Q-Q plot and box plot (2.698 σ). Attention was aimed at outlying observations (outlying proteins) suggested by boxplot which are those that appear to deviate markedly from other members of the data set in which they occur [56]. All differently accumulated proteins including outliers were then sorted into clusters according to their mode of accumulation using hierarchical clustering (package "stats", function "hclust", Ward's method). Dendrograms were cut at height of $h = 10$ so that clusters with distinctly different modes of accumulation were created. Furthermore, the changes in protein abundance were displayed using Heat map (packages "gplots" and "RColorBrewer", function "heatmap.2"). Since the proteins differed greatly in their abundances, all data were normalized (row centered and scaled).

4. CONCLUSIONS

In this study, processes of winter barley cold hardening have been glimpsed at proteome level by monitoring and comparing the response of leaves and crowns to various periods of cold and frost. It has been observed that protein and nitrogen metabolic processes were influenced by low temperatures to a similar extent in both tissues while catabolism, carbohydrate metabolism and stress response were more affected in crowns. We believe that these differences may reflect the fact that crowns are crucial for whole plant survival. Amongst the proteins which showed large changes in abundance, and therefore might play substantial roles in frost tolerant winter barley response to low temperature, AAA ATPase in leaves or HSP70 in crowns have been found. Chloroplast proteins were frequently observed as differently accumulated. Therefore, our results denote the importance of chloroplasts in low temperature response of barley. All proteins which significantly (at least twice) changed their accumulation in course of cold or frost periods were assorted to clusters according to their accumulation profile; there have been found five protein clusters in leaves and three clusters in crowns. We concluded that these clusters are likely to have a decisive influence in distinct phases of barley cold hardening as the proteins in different clusters mounted their abundance in diverse time points of low temperature exposure.

The molecular basis of winter cereals cold hardening process has been so far studied predominantly by transcriptomic methods. Generally, our proteomic study showed an accordance with previous transcriptomic studies; nevertheless, there may exist substantial differences between mRNA expression and actual protein levels due to different mechanisms of gene expression regulation. Therefore, further proteomic studies are needed in order to understand cold response regulation in cereals more profoundly.

The presented results can be considered as a starting point in selection process of frost resistance marker proteins suitable for effortless and fast crop improvement. In this perspective, it will be necessary to carry out follow-up studies to compare more barley genotypes differing in their frost resistance; from this point of view, our experiment can be regarded as a pilot study.

ACKNOWLEDGMENTS

This research was funded by the Ministry of Agriculture of the Czech Republic, projects QH 81287, QJ1310055 and MZE0002700604. We are obliged to Anna Janská for providing plant samples and unknown reviewers of their helpful remarks on our manuscript.

REFERENCES

1. Newton, A.C.; Flavell, A.J.; George, T.S.; Leat, P.; Mullholland, B.; Ramsay, L.; Revoredo-Giha, C.; Russell, J.; Steffenson, B.J.; Swanston, J.S.; et al. Crops that feed the world 4. Barley: A resilient crop? Strengths and weaknesses in the context of food security. Food Sec **2011**, 3, 141–178.
2. Limin, A.E.; Fowler, D.B. Developmental traits affecting low-temperature tolerance response in near-isogenic lines for the vernalization locus Vrn-A1 in wheat (Triticum. aestivum L. em Thell). Ann. Bot **2002**, 89, 579–585.
3. Hommo, L. Hardening of some winter wheat (Triticum. aestivum L.), rye (Secale. cereale L.), triticale (× Triticosecale Wittmack) and winter

barley (Hordeum. vulgare L.) cultivars during autumn and the final winter survival in Finland. Plant Breed **1994**, 112, 285–293.

4. Kacperska, A. Plant Response to Low Temperature: Signaling Pathways Involved in Plant Acclimation. In Cold-Adapted Organisms—Ecology, Physiology, Enzymology and Molecular Biology; Margesin, R., Schinner, F., Eds.; Springer: Berlin, Germany, 1999; pp. 79–103.
5. Thomashow, M.F. Plant cold acclimation: Freezing tolerance genes and regulatory mechanisms. Annu. Rev. Plant Physiol. Plant Mol. Biol **1999**, 50, 571–599.
6. Timperio, A.M.; Egidi, M.G.; Zolla, L. Proteomics applied on plant abiotic stresses: Role of heat shock proteins (HSP). J. Proteomics **2008**, 71, 391–411.
7. Kosová, K.; Vítámvás, P.; Prášil, I.T. Expression of dehydrins in wheat and barley under different temperatures. Plant Sci **2011**, 180, 46–52.
8. Kamal, A.H.M.; Kim, K.H.; Shin, K.H.; Choi, J.S.; Baik, B.K.; Tsujimoto, H.; Heo, H.Y.; Park, C.S.; Woo, S.H. Abiotic stress responsive proteins of wheat grain determined using proteomics technique. Aust. J. Crop. Sci **2010**, 4, 196–208.
9. Sarhadi, E.; Mahfoozi, S.; Hosseini, S.A.; Salekdeh, G.H. Cold acclimation proteome analysis reveals close link between the up-regulation of low-temperature associated proteins and vernalization fulfillment. J. Proteome Res **2010**, 9, 5658–5667.
10. Rinalducci, S.; Egidi, M.G.; Karimzadeh, G.; Jazii, F.R.; Zolla, L. Proteomic analysis of a spring wheat cultivar in response to prolonged cold stress. Electrophoresis **2011**, 32, 1807–1818.
11. Vítámvás, P.; Prášil, I.T.; Kosová, K.; Planchon, S.; Renaut, J. Analysis of proteome and frost tolerance in chromosome 5A and 5B reciprocal substitution lines between two winter wheats during long-term cold acclimation. Proteomics **2012**, 12, 68–85.
12. Fowler, D.B.; Breton, G.; Limin, A.E.; Mahfoozi, S.; Sarhan, F. Photoperiod and temperature interactions regulate low-temperature-induced gene expression in Barley. Plant Physiol **2001**, 127, 1676–1681.
13. Kosová, K.; Holková, L.; Prášil, I.T.; Prášilová, P.; Bradáčová, M.; Vítámvás, P.; Čapková, V. Expression of dehydrin 5 during the development of frost tolerance in barley (Hordeum. vulgare L.). J. Plant Physiol **2008**, 165, 1142–1151.

14. Vítámvás, P.; Saalbach, G.; Prášil, I.T.; Čapková, V.; Opatrná, J.; Ahmed, J. WCS120 protein family and proteins soluble upon boiling in cold-acclimated winter wheat. J. Plant Physiol **2007**, 9, 1197–1207.
15. Vítámvás, P.; Prášil, I.T. WCS120 protein family and frost tolerance during cold acclimation, deacclimation and reacclimation of winter wheat. Plant Physiol. Biochem **2008**, 46, 970–976.
16. Greenup, A.G.; Sasani, S.; Oliver, S.N.; Walford, S.A.; Millar, A.A.; Trevaskis, B. Transcriptome analysis of the vernalization response in barley (Hordeum. vulgare L.) seedlings. PLoS One **2011**, 6, e17900.
17. Svensson, J.T.; Crosatti, C.; Campoli, C.; Bassi, R.; Stanca, A.M.; Close, T.J.; Cattivelli, L. Transcriptome analysis of cold acclimation in barley Albina and Xantha Mutants. Plant Physiol **2006**, 141, 257–270.
18. Armbruster, U.; Pesaresi, P.; Pribil, M.; Hertle, A.; Leister, D. Update on chloroplast research: New tools, new topics and new trends. Mol. Plant **2001**, 4, 1–16.
19. Šimková, K.; Kim, C.; Gacek, K.; Baruah, A.; Laloi, C.; Apel, K. The chloroplast division mutant caa33 of Arabidopsis thaliana reveals the crucial impact of chloroplast homeostasis on stress acclimation and retrograde plastid-to-nucleus signaling. Plant J **2012**, 69, 701–712.
20. Tanino, K.K.; McKersie, B.D. Injury within the crown of winter wheat seedlings after freezing and icing stress. Can. J. Bot **1985**, 63, 432–436.
21. Livingston, D.P.; Premakumar, R.; Tallury, S.P. Carbohydrate partitioning between upper and lower regions of the crown in oat and rye during cold acclimation and freezing. Cryobiology **2006**, 52, 200–208.
22. Winfield, M.O.; Lu, C.; Wilson, I.D.; Coghill, J.A.; Edwards, K.J. Plant responses to cold: Transcriptome analysis of wheat. Plant Biotechnol. J **2010**, 8, 749–771.
23. Janská, A.; Aprile, A.; Zámečník, J.; Cattivelli, L.; Ovesná, J. Transcriptional responses of winter barley to cold indicate nucleosome remodelling as a specific feature of crown tissues. Funct. Integr. Genomics **2011**, 11, 307–325.
24. Renaut, J.; Lutts, S.; Hoffmann, L.; Hausman, J.F. Responses of poplar to chilling temperatures: Proteomic and physiological aspects. Plant Biol **2004**, 6, 81–90.
25. Hashimoto, M.; Komatsu, S. Proteomic analysis of rice seedlings during cold stress. Proteomics **2007**, 7, 1293–302.

26. Ganeshan, S.; Vítámvás, P.; Fowler, D.B.; Chibbar, R.N. Quantitative expression analysis of selected COR genes reveals their differential expression in leaf and crown tissues of wheat (Triticum aestivum L.) during an extended low temperature acclimation regimen. J. Exp. Bot **2008**, 59, 2393–2402.
27. Limin, A.E.; Corey, A.; Hayes, P.; Fowler, D.B. Low-Temperature acclimation of barley cultivars used as parents in mapping populations: Response to photoperiod, vernalization and phenological development. Planta **2007**, 226, 139–146.
28. Kjellsen, T.D.; Shiryaeva, L.; Schröder, W.P.; Strimbeck, G.R. Proteomics of extreme freezing tolerance in Siberian spruce (Picea. obovata). J. Proteomics **2010**, 73, 965–975.
29. Ifuku, K.; Ishihara, S.; Sato, F. Molecular functions of oxygen-evolving complex family proteins in photosynthetic electron flow. J. Integr. Plant Biol **2010**, 52, 723–734.
30. Cui, S.; Huang, F.; Wang, J.; Ma, X.; Cheng, Y.; Liu, J. A proteomic analysis of cold stress responses in rice seedlings. Proteomics **2005**, 5, 3162–3172.
31. Schweighofer, A.; Hirt, H.; Meskiene, I. Plant PP2C phosphatases: Emerging functions in stress signaling. Trends Plant Sci **2004**, 9, 236–243.
32. Tahtiharju, S.; Palva, T. Antisense inhibition of protein phosphatase 2C accelerates cold acclimation in Arabidopsis thaliana. Plant J **2001**, 26, 461–470.

33. Frydman, J. Folding of newly translated proteins in vivo: The role of molecular chaperones. Annu. Rev. Biochem **2001**, 70, 603–647.
34. Straus, M.R.; Rietz, S.; ver Loren van Themaat, E.; Bartsch, M.; Parker, J.E. Salicylic acid antagonism of EDS1-driven cell death is important for immune and oxidative stress responses in Arabidopsis. Plant J **2010**, 62, 628–640.
35. Vítámvás, P.; Kosová, K.; Prášilová, P.; Prášil, I.T. Accumulation of WCS120 protein in wheat cultivars grown at 9 C or 17 C in relation to their winter survival. Plant Breed **2010**, 129, 611–616.
36. Oikawa, A.; Nagai, K.; Kato, K.; Kidou, S. Gene silencing of barley P23k involved in secondary wall formation causes abnormal tiller formation and intercalary elongation. Breed. Sci **2009**, 59, 664–670.

37. Lehmann, J.; Atzorn, R.; Bruckner, C.; Reinbothe, S.; Leopold, J.; Wasternack, C.; Parthier, B. Accumulation of jasmonate, abscisic-acid, specific transcripts and proteins in osmotically stressed barley leaf segmants. Planta **1995**, 197, 156–162.
38. Dunn, M.A.; Morris, A.; Jack, P.L.; Hughes, M.A. A low-temperature-responsive translation elongation factor lα from barley (Hordeum. vulgare L.). Plant Mol. Biol **1993**, 23, 221–225.
39. Wallsgrove, R.M.; Turner, J.C.; Hall, N.P.; Kendall, A.C; Bright, S.W. Barley mutants lacking chloroplast glutamine synthetase-biochemical and genetic analysis. Plant Physiol **1987**, 83, 155–158.
40. Pang, Q.Y.; Chen, S.X.; Dai, S.J.; Chen, Y.Z.; Wang, Y.; Yan, X.F. Comparative proteomics of salt tolerance in Arabidopsis thaliana and Thellungiella. halophila. J. Proteome Res **2010**, 9, 2584–2599.
41. Yan, S.P.; Tang, Z.C.; Su, W.; Sun, W.N. Proteomic analysis of salt stress-responsive proteins in rice root. Proteomics **2005**, 5, 235–244.
42. Cai, H.; Zhou, Y.; Xiao, J.; Li, X.; Zhang, Q.; Lian, X. Overexpressed glutamine synthetase gene modifies nitrogen metabolism and abiotic stress responses in rice. Plant Cell Rep **2009**, 28, 527–537.
43. Skinner, D.Z. Post-Acclimation transcriptome adjustment is a major factor in freezing tolerance of winter wheat. Funct. Integr. Genomics **2009**, 9, 513–523.
44. Hanano, A.; Burcklen, M.; Flenet, M.; Ivancich, A.; Louwagie, M.; Garin, J.; Blee, E. Plant seed peroxygenase is an original heme-oxygenase with an EF-hand calcium binding motif. J. Biol. Chem **2006**, 281, 33140–33151.
45. Thangasamy, S.; Chen, P.W.; Lai, M.H.; Chen, J.; Jauh, G.Y. Rice LGD1 containing RNA binding activity affects growth and development through alternative promoters. Plant J **2012**, 71, 288–302.
46. Liu, X.D.; Shen, Y.G. NaCl-Induced phosphorylation of light harvesting chlorophyll a/b proteins in thylakoid membranes from the halotolerant green alga, Dunaliella. salina. FEBS Lett **2004**, 569, 337–340.
47. Wang, W.; Tai, F.J.; Chen, S.N. Optimizing protein extraction from plant tissues for enhanced proteomics analysis. J. Sep. Sci **2008**, 31, 2032–2039.
48. Strohalm, M.; Kavan, D.; Novak, P.; Volny, M.; Havlicek, V mMass 3: A Cross–Platform software environment for precise analysis of mass spectrometric data. Anal. Chem **2010**, 11, 4648–4651.
49. Barley1.na32.annot.csv. Available online: http://www.affymetrix.com/estore/ (accessed on 3 February 2013).

50. UniProtKB. Available online: http://www.uniprot.org/ (accessed on 3 February 2013).
51. The Gene Ontology Consortium. Gene ontology: Tool for the unification of biology. Nat. Genet **2000**, 1, 25–29.
52. Emanuelsson, O.; Nielsen, H.; Brunak, S.; von Heijne, G. Predicting subcellular localization of proteins based on their N-terminal amino acid sequence. J. Mol. Biol **2000**, 300, 1005–1016.
53. Ye, J.; Fang, L.; Zheng, H.K.; Zhang, Y.; Chen, J.; Zhang, Z.J.; Wang, J.; Li, S.T.; Li, R.Q.; Bolund, L.; et al. WEGO: A web tool for plotting GO annotations. Nucleic Acids Res **2006**, 34, W293–W297.
54. WEGO. Available online: http://wego.genomics.org.cn/cgi-bin/wego/index.pl (accessed on 1 February 2013).
55. R 2.15.0. Available online: http://www.r-project.org/ (accessed on 1 October 2012).

13

CHAPTER

PROPERTIES OF SOYBEAN FOR BEST POSTHARVEST OPTIONS

Seth I. Manuwa

Department of Agricultural Engineering, School of Engineering and Engineering Technology, The Federal University of Technology, Akure Nigeria

INTRODUCTION

Soybean is considered as one very important grain grown commercially in more than 35 countries of the world and the leading producer is the USA (41%) followed by Brazil (23%), Argentina (16%) and China (9%), (F A O 1988).

Soybean contains 40% protein, 35% total carbohydrate and 20% cholesterol-free oil (Deshpande et al., 1993). Mineral content of whole soybean is about 1.7% for potassium, 0.3% for Magnesium, 110 ppm iron, 50 ppm zinc and 20 ppm copper (Smith and Circle, 1972). Soybean is the world leading vegetable oil and accounts for about 20 to 24% of all fats and oil in the world. Soybean is becoming increasingly important in agriculture because it is a food source in human and animal nutrition

So many varieties have been developed around the world considering desired traits. The properties of the developed cultivars could be considered to vary from one cultivar to the other. Sometimes, such variations in properties (especially physical properties) are easily observable, especially in the size and shape of such cultivars. Other properties would have to be measured to know them or to see how they vary from one cultivar to another. By extension, the properties (physical, mechanical and chemical) of a cultivar affect the post harvest options to which a cultivar may be subjected. The challenge of post harvest processing of soybean into animal and human food is increasing by the day. This is so because, the world's population is increasing and the challenge of eradicating hunger and producing quality food on the surface of the earth is staring.

Manuwa (2000, 2007), Manuwa et al.(2004, 2005) reported on similar improved varieties of Soybean that were developed in Nigeria. The major improvements made on soybean varieties from 1987 through 1992 at IITA were to increase grain yield by about 20%, improve resistance to pod shattering and to maintain the level of all other traits constant. In order to design equipment for threshing, winnowing, separation, grading, sorting, size reduction, storage, and other secondary processing of soybean, especially the new improved cultivars, the physical properties should be determined.

VARIETIES OF SOYBEANS

So many varieties of soybean have been developed around the world so that it is a major task to know all of them. The main aim of developing varieties (cultivars) was to improve desired traits such as:

- Early maturity,
- Disease resistance e.g phytophthora root rot resistant
- High grain yield,
- Shattering and lodging resistant,
- Intact seed coat and some weathering tolerance,
- Seed quality that meets culinary market standards, for example a light hila culinary type.

A number of varieties have been reported in literature (Tables 1, 2, 3).

HARVESTING AND UTILISATION OF SOYBEAN

Needless to say that before soybean can be utilised as food for either man or animal, it must first of all be harvested from the field. However, harvest management is a crucial skill for the specialty soybean producer, simply because the physical appearance of the beans is so important to the buyer. Small-seeded soybeans tend to thresh well, but air adjustments may have to be fine-tuned to remove chaff without blowing the small seeds out the back of the combine. Large-seeded soybeans are extremely prone to mechanical damage during threshing operations, which can knock off the seed coat and/or split the embryo into its cotyledonal halves. The combine's cylinder speeds will have to be slowed considerably to avoid this, and the crop may require harvesting at somewhat higher moisture content.

Prompt harvesting will always be a must, as field deterioration of the seed affecting appearance can commence soon after the moisture content of the physiologically mature seed drops to 14%.. If storage is necessary, the producer will have to ensure that storage facilities are clean, dry, and free from any materials that may be toxic to humans. The conditions under which beans are stored greatly influence the quality of the processed product. Moisture content of 13% or less will prevent mold growth. However, very dry beans tend to split when being transferred, and the splitting lowers the quality.

Soybeans can be used for oil, livestock feeds and for preparing various dishes. A number of traditional foods have been produced from soybeans: *Tofu. Miso, Natto, Tempeh, Soymilk, Soyflour, Soyoil, soy milk* (Bschmann, 2001). According to the report, the size of the seed is often crucial, and may be either smaller or larger than average soybean cultivars. For example, small seeds are sought out for *natto,* while large seeds are preferred for *tofu.* Perfectly round seeds are generally prized, while oblong or kidney-shaped soybeans are usually avoided

POST HARVEST OPTIONS

Post harvest options are generally all the activities that can be carried out after the harvesting of crops in order to convert it to use by man and animal. It can be classified into primary and secondary processing.

Primary processing: This includes threshing, winnowing, cleaning, separation, grading, sorting, packaging, transportation, marketing, storage and so on.

Grains or seeds from harvesters are not directly suitable for its final use such as re-sowing, animal feed or human consumption. The standards of seeds in the three categories have risen in the last few decades to date. Reasons, especially for re-sowing seeds include the need to achieve international marketing standard, and secondly the uniform, high germination product required in precision drilling.

Table 1. Some Soybean cultivars from USA, Australia, Slovenia & Nigeria.

COUNTRY	VARIETY	YIELD (Kg/ha)	SOURCE
USA	Jim	-	www.ag.ndsu.nodak.edu/aginfo/variety/soybean.htm
	Traill	-	
	RG200RR	-	
	Walsh	-	
	MN0201	-	
	MN0302	-	
	Barnes	-	
	Normatto	-	
	Nannonatto	-	
	Norpro	-	
	SD1081RR	-	
	Sargent	-	
	Surge	-	
	SD1091RR	-	
AUSTRALIA	Arunta	3.81	Adapted from: www.ag.ndsu.nodak.edu/aginfo/variety/ Soybean.
	Stephens	3.80	
	Bowyer	3.55	
	Curringa	3.73	
	Djakal (BAF 212)	3.93	
SLOVENIA	Aldama	1791	Acko and Trdan (2009)
	Borostyan	1242	
	Essor	2757	
	Ika	3138	
	Kador	3702	
	Major	2342	
	Nawiko	2748	
	Olna	2272	
	Tarna	3381	
	Tisa	4216	
NIGERIA	Samsoy 2	1745	Manuwa, 2005; 2007
	TGx 923-2E	1736	
	TGx992-22E	1642	
	TGx 1440-1E	1629	
	TGx 1448-2E	1558	
	TGx 1660- 19F	2134	
	TGx 1489-1D	2071	
	TGx 1447-2D	1970	
	TGx 1437-1D	1877	
	TGx 1455- 2E	1660	
	TGx 849-313D	1524	

Table 2. Some Soybean cultivars from India.

COUNTRY	VARIETY	YIELD (Kg/ha)	OIL CONTENT (%)
INDIA	Alankar	2200	-
	Ankur	2300	-
	Clark - 63	1800	-
	PK-1042	3300	-
	PK-262	2800	-
	PK-308	2600	20-23
	PK-327	2300	-
	PK-416	3200-3800	41-56
	PK-564	3000	-
	Shilajeeth	2200	-
	Bragg	1800	-
	Calitur	1800	-
	Durga	2100	-
	Durga	2100	-
	Gaurav	2200	-
	Indira Soya -9	2300	-
	JS-2	1800	-
	JS-71-05	2000-2400	41
	JS-75-46	1600-3100	-
	JS-76-205	1600-2000	-
	JS-79-81	2800	-
	JS-80-21	2500-3000	-
	JS-90-41	2500-3000	-
	JS-335	2500-3000	17-19
	MACS-13	2700	15-22
	MACS-58	2000-2500	-
	MAUS-47 (Parbhani ona)	2500-3000	20
	MS-335	2800	-
	NRC-12(Ahilya-2)	2800	-
	NRC-2(Ahilya-1)	3500-4000	21
	NRC-7(Ahilya-3)	3200	-
	PK-472	3300	-
	PUSA-16	2800	-
	PUSA-22	2600	-
	PUSA-37	2800	-
	TYPE-49	2200	-
	MACS-57	2800	-
	MACS-450	2500	20
	MAUS-2	2450	-
	MAUS-1	2800	-
	MAUS-32(Prasad)	3000-3500	19
	KB-79(Sneha)	1700	-
	MACS-124	2500-3200	-
	PUSA-40	2600	-

Table 3. Some Soybean cultivars from Alabama, USA.

BR 16	BR 36	BRS 153	BRS 155	Embrapa 1
Embrapa 48	FT 106 I	FT 109 I	FT 2	FT 20 (Jau)
FT 4	FT 7 (Taroba)	FT 9 (Inae)	FT Manaca	FT Seriema
IAC 13	IAC 15	IAC 15-1	IAC 16	IAC 4
IAC Foscarin-31	IAC/Holambra twart-1	KI-S 601	KI-S 602 RCH	MS/BR 34 (Empaer 10)
Ocepar 10	Ocepar 16	Ocepar 4 (Iguaçu)	Ocepar 7 (Brilhante)	Ocepar 8
RB 502	RS 9 (Itaúba)	BRS 156	IAC 11	Paraná
BRS 157	BRSMS Apaiari	CEP 12 (Cambará)	Cobb	FT 103
FT 104	FT 2000	IAS 4	Ivorá	Ocepar 17
Ocepar 5 (Piquiri)	RS 5 (Esmeralda)	BRS 134	BRS 136	BRS 138
BRS 65	BRS 66	BRSMA Sambaíba	BRSMA Seridó RCH	BRSMG Confiança
BRSMS Piapara	BRSMS Piracanjuba	CEP 20 (Guajuvira)	DM Nobre	Embrapa 30 (V. R Doce)
Embrapa 62	Emgopa 313 (Anhang.)	Emgopa 316 (Rio Verde)	FT 101	FT 19 (Macacha)
GO/BR 25 (Aruanã)	IAC 100	IAC 12	MS/BR 19 (Pequi)	Ocepar 14
Santa Rosa	BR 28 (Seridó)	BR 38	BRS Carla	RB 603
RB 604	DM 247	DM 339	1 BR 6 (Nova Bragg) Bragg(3)	Bragg
BRS 137	BRS 154	BRS Celeste	BRSMG Garantia	BRSMG Robusta
BRSMG Segurança	BRSMG Virtuosa	BRSMS Mandi	Embrapa 20	(Doko RC)
Embrapa 63 (Mirador)	Emgopa 315 (R. Verm.)	FT 10 (Princesa)	FT 18 (Xavante	FT 6 (Veneza
FT Cometa	IAC 18	IAC 22	MG/BR48 (Gar. RCH)	
UFV 19 FT	UFV/ITM-1	BR 30	BRS 135	BRS Milena
BRSMS Carandá	BRSMS Lambari	BRSMS Piraputanga	BRSMS Taquari	BRSMS Tuiuiú
DM Soberana	Embrapa 64 (Ponta Porã)	Emgopa 301	FT 14 (Piracema)	FT 5 (Formosa)
FT Abyara	FT Maracajú	FT Saray	Fundacep 33	Ocepar 12
UFV 10 (Uberaba)	Bossier	BR IAC 21	BRSMA Parnaíba	BRSMG 68 (Vencedora)
BRSMG Liderança	BRSMG Renascença	BRSMS Bacuri	BRSMS Surubi	DM Vitória
FT 11 (Alvorada)	FT Guaira	IAC 17	IAC 8	IAC 8-2
KI-S 702	KI-S 801	MG/BR-46 (Conquista)	Ocepar 3 (Primavera)	UFV 18 (Patos de Minas)
BRSGO Goiatuba	BR 4	BR 9 (Savana)	BRSMA Pati	Embrapa 4
Embrapa 46	Embrapa 47	Emgopa 304 (Campeira)	Emgopa 309 (Goiana)	FT 8 (Araucária)
FT Bahia	FT Cristalina	FT Cristalina CH	FT Estrela	FT Iramaia
FT Líder	Ivaí	MT/BR 50 (Parecis)	MT/BR 51 (Xingu)	MT/BR 53 (Tucano)
Planalto	UFV 5	BRSMT Crixás	CAC-1	CS 301
CS 303	DM 118	Dourados	FEPAGRO-RS 10	FT 102
IAC 20	M-SOY 2002	BRSGO Catalão	Campos Gerais	Embrapa 9 (Bays)
Emgopa 308 (S.Dourada)	FT 100	FT 45263	FT Canarana	FT Eureka
IAC 14	Invicta	Ipagro 21	RS 7 (Jacuí)	Emgopa 303

Requirements for seed cleaning:

- To obtain graded lots of seed which will meet home and international testing standards for the variety under consideration, in terms of purity, viability, vigour and size variation
- To remove completely any seeds, the sale of which in a batch may contravene the Noxious Weeds Act, of some countries.
- To avoid any loss of good seeds in the cleaning process.
- To avoid excessive wear that may be due to sorting machines.
- To remove all contaminants that is capable of damaging subsequent processing machinery such as size reduction machines. Typical contaminants include weed seeds, straws, leaves, stones and soil particles.

Principles of separation:

It is important to identify differences in the physical properties of the seeds and the contaminants that will enable the machine (to be designed) make them flow in different directions. Such properties include the following:

- Seed dimensions: length, width, thickness, geometric mean diameter
- Specific gravity
- Falling rate (float)
- Surface texture, friction
- Colour
- Resilience (ability to bounce)
- Electrical conductance

Typically, most processing machines identify differences in properties between good seeds and contaminants. For example a sieve identifies size while other machines identify a combination of properties such as specific gravity table. The shaking table for example identifies friction, size and density of the seeds. The air-screen cleaner for example make use of differences in size, shape and density of the seeds and such machine range from a small, one fan, single screen machine to the large multi-fan eight screen machine with several air columns. Other machines that are used for primary processing include threshing machine, from simple hand operated threshers to high capacity multi cop threshers, combine harvesters, winnowers, air-screen separators (oscillating or vibrating), graders (band, spiral), separators (spiral, table, magnetic, electrostatic, colour, pneumatic, and so on).

Secondary crop processing: It involves processing of food for direct consumption. This requires grinding, milling, oil extraction and so on. To accomplish these, machines are used such as size reduction machines such as milling machines, dehullers, grinding machines, oil press and so on.

METHODOLOGY FOR EVALUATING SOYBEAN PROPERTIES

Sample preparation: Dry mature Soybeans [Glycine max.] are normally used for all the experiments. Before the experiments, the grains were further cleaned by removing those that were physically bad, unhealthy or broken. The moisture content of the grain would be determined using a standard method. Physical properties were determined at the initial moisture content. Thereafter, grain sample of the desired moisture levels were prepared by adding calculated amount of distilled water and sealed in separated polythene bags. The samples would be kept at about 278 ^{0}K in a refrigerator for 1 week to enable the moisture to distribute uniformly throughout the sample. Before the commencement of a test, the required quantity of the grain was taken out of the refrigerator (if kept there to cool), and allowed to warm up to room temperature at about 305 ^{0}K.

Physical properties: The physical properties of Soybean to be determined include linear dimensions, mass, bulk density, seed density, volume, surface area, sphericity, porosity, coefficient of static friction on structural surfaces and angle of repose, angle of internal friction, terminal velocity. Experiments were conducted at five levels of moisture content in the desired range and replicated five times. Average values were normally reported. The choice of the range of moisture content was due to the fact that the lower limit was the safe storage moisture content, and the upper range, the maximum moisture content obtainable after the seeds were soaked overnight.

Linear dimensions and geometric mean diameter: To determine the size of the grain, 10 sub samples each consisting of 100 grains were randomly taken. From each sub sample, 10 grains were taken and their three linear dimensions namely, length (L), width (W) and thickness (T) were measured with a venier calipers having accuracy of 0.01mm. The geometric mean diameter (DGM) of the grain was calculated by using the following relationship (Sreenarayanan et al 1985, Sharma et al 1985).

$$D_{GM} = (LWT)^{1/3} \tag{1}$$

Test weight: Sub samples of One, one hundred and one thousand soybean grains from each sample were randomly selected and weighed. The averages of the replicated values are usually reported.

Bulk and seed density: A method similar to that reported by Shephered and Bhardwaj (1986) can be used to determine the bulk density at each moisture level: a 180 ml cylinder was filled continuously from a height of about 15 cm. Tapping during filling was done to obtain uniform packing and to minimize the wall effect, if any. The filled sample was weighed and the bulk density of the material filling the cylinder was computed (Shephered and Bhardwaj, 1986; Deshpande and Ali, 1988; Mohsenin, 1970). The seed density of the grain can be determined by the liquid displacement method to determine the seed volume similar to that reported (Shephered and Bhardwaj, 1986; Deshpande and Ali, 1988).

Sphericity and porosity: According to Mohsenin (1970), sphericity ϕ, was calculated using the formula.

$$\phi = \frac{(LWT)^{1/3}}{L} \tag{2}$$

Fractional porosity is defined as the fraction of space in the bulk grain which is not occupied by grain. Thompson and Isaacs (1967) gave the following relationship for fractional porosity.

$$\varepsilon = \frac{(1-\rho_b)}{\rho_s} x100 \tag{3}$$

Angle of repose: The emptying angle of repose θ is normally determined at the moisture levels using the pipe method (Henderson 1982, Jha 1999). A pipe of 40 cm height and 106 mm internal diameter was kept on the floor vertically and filled with the sample, Tapping during filling was done to obtain uniform packing. The tube was slowly raised above the floor so that the whole material could slide and form a heap. The height above the floor H and the diameter of the heap D at its base were measured with a measuring scale and the angle of reposes θ of the soybean computer using the equation.

$$\theta = Arc\tan(2H / D) \tag{4}$$

Surface area: The surface area of the grain can be found by analogy with a sphere of geometric mean diameter for the different levels given by (McCabe et al., 1986)

$$S = \pi D_{GM}^{2} \tag{5}$$

Coefficient of static friction: The coefficient of static friction for seed grain can be determined against structural surfaces such as plywood (with grain parallel to direction of motion and then with grain perpendicular to direction of motion), galvanized steel (GS), glass, concrete and so on. A bottom less wooden box of 150 mm x 150 mm x 40 mm was constructed for this purpose. This was similar to that reported by (Oje, 1994). The box shall be filled with soybean grains on an adjustable tilting surface. The surface would be raised gradually using a screw device until the box started to slide down and the angle of inclination read on a graduated scale.

Terminal velocity: The terminal velocity of soybean at different moisture content can be determined using an air column (Polat et al., 2006). For each test, a seed was dropped from the top of a 75 mm diameter, 1 m long glass tube. The air was made to flow upwards in the tube from bottom to the top and the air velocity at which the sample seed was suspended was noted with an anemometer having at least 0.1 m/s sensitivity.

Angle of internal friction: To determine the angle of internal friction of soybean at different moisture contents, the direct shear method can be used according to Uzuner (1996), Zou and Brucewitz (2001), Molenda et al.(2002) and Mani et al.(2004). Typical velocity to be used during the experiment is 0.7 mm/min (Kibar and Ozturk, 2008) and the angle of internal friction can be calculated using the following equations:

$$\sigma = \frac{N}{A} 100 \tag{6}$$

$$\tau = \frac{T_s}{A} 100 \tag{7}$$

$$\tau = (C + \sigma tg\phi) \tag{8}$$

Table 4. Effect of moisture content on mass and dimensional properties of some soybean cultivars.

Cultivars	MC range %(db)	Dimensions (mm)	Mass (g)	Reference
JS- 7244	8.7- 25.0	L: 6.32 to 6.75 W: 5.23 to 5.55 T: 3.99 to 4.45 GMD: 5.09 to 5.51	- - 1000 grains: 110 to 127	Deshpande et al., 1993
TGX 1440-1E	10.5- 34.1	L: 8.58 to 10.02 W: 6.51 to 7.22 T: 5.43 to 5.69 GMD: 6.71 to 7.44	1 grain: 0.11 to 0.21 100 grains: 14.67 to 19.98 1000 grains: 139.18 to 190.6	Manuwa, 2000
TGX 1871- 5E	7.1- 43.7	L: 7.52- 9.11 W: 6.47- 7.05 T: 5.49- 5.05 GMD: 6.44- 7.29	1 grain: 0.136 to 0.205 100 grains: 12.3 to 16.59 1000 grains: 119.17 to 153.15	Manuwa and Afuye, 2004
TGX 1019-2EB	6.7- 47.1	L: 7.37 to 9.96 W: 6.48 to 7.45 T: 5.33 to 5.54 GMD: 6.33 to 7.39	1 grain: 0.178 to 0.213 100 grains: 13.78 to 18.79 1000 grains: 130.67 to 180.21	Manuwa and Odubanjo, 2005
Unspecified	6.7- 15.3	L: 7.41 to 9.57 W: 5.34 to 6.75 T: 4.5 to 5.17 GMD: 5.62 to 6.94	1000 grains: 121.76 to 223.65	Polat et al., 2006
TGX 1448- 2E	9.9 to 39.6	L: 8.3 to 10.4 W: 6.4 to 7.5 T: 5.4 to 5.8 GMD: 6.6 to 7.6	1 grain: 0.19 to 0.24 100 grains: 15.6 to 19.4 1000 grains: 154.2 to 185.6	Manuwa, 2007
Unspecified	8- 16	L: 7.24 to 8.19 W: 6.79 to 7.12 T: 5.78 to 6.23 GMD: 6.57 to 7.14	NAV	Kibar and Ozturk, 2008
Unspecified	6.92- 21.19	L: 7.27 to 8.23 W: 6.48 to 6.97 T: 5.41 to 5.94 GMD: 6.34 to 6.98	1000 grains: 171.5 to 219.04	Tavakoli et al., 2009

MC= moisture content, NAV= not available

RUPTURE FORCE AND RUPTURE ENERGY

To determine the rupture force and rupture energy, a Universal Testing Machine (UTM) can be used such as Instron Universal Testing Machine reported by Tavakoli et al. (2009). It was equipped with a 500 kg compression load cell and integrator. The measurement accuracy was 0.001 N in force and 0.001 mm in deformation. The individual grain was loaded between two parallel plates of the machine and compressed along with thickness until rupture occurred as is denoted by a rupture point in the force-deformation curve. The rupture point is a point on the force-deformation curve at which the loaded specimen shows a visible or invisible failure in the form of breaks or cracks. According to them the point was detected by a continuous decrease of the load in the force-deformation diagram. The loading rate of 5 mm/min was used according to ASAE (2006a). The energy absorbed by the sample at rupture was determined by calculating the area under the force-deformation curve from the relationship:

$$E_a = \frac{F_r D_r}{2} \tag{9}$$

Where Ea is the rupture energy in mJ, Fr is the rupture force in N and Dr is the deformation at rupture point (Braga et al., 1999).

Table 5. Effect of moisture content on density, sphericity, porosity and terminal velocity of some soybean cultivars.

Cultivars	MC range %(db)	Seed density	Bulk density	Sphericity (%)	Porosity (%)	V_t (m/s)	Reference
JS- 7244	8.7- 25.0	1216 - 1124	735 - 708	80.6 – 81.6	40 - 37	NAV	Deshpande et al., 1993
TGX 1440-1E	10.5- 34.1	1184 - 1076	720 - 631	79 – 73.3	23.6 – 34.2	NAV	Manuwa, 2000
TGX 1871- 5E	7.1- 43.7	1222.3 – 935.7	686.5 – 616.7	85.87- 78.23	25.64 – 40.96	NAV	Manuwa and Afuye, 2004
TGX 1019-2EB	6.7- 47.1	1157 - 952	728.5 – 608.4	86 – 74.9	23.46 – 42.33	NAV	Manuwa and Odubanjo, 2005
Unspecified	6.7- 15.3	1062.6 to 1086.5	804.8 to 689.3	75 to 72	51 to 44.2	7.13 to 9.24	Polat et al., 2006
TGX 1448- 2E	9.9 to 39.6	1465 - 1074	714 - 638	79.1 – 72.7	19.5 – 33.7	NAV	Manuwa, 2007
Unspecified	8- 16	983.33 – 905.67	766.12 – 719.00	91- 87	22.58 – 20.61	NAV	Kibar and Ozturk, 2008
Unspecified	6.92- 21.19	1147.86 to 1126.43	650.95 to 625.36	87.25 to 84.75	43.29- 44.48	NAV	Tavakoli et al., 2009

MC= moisture content, NAV= not available
Vt = terminal velocity

Table 6. Effect of moisture content on angle of repose and coefficient of static friction of some soybean cultivars.

Cultivars	MC range %(db)	Angle of repose (degree)	Coefficient of static friction				Reference
			Galvanised steel	PWLG	PWDG	Glass	
JS- 7244	8.7- 25.0						Desahpande et al., 1993
TGX 1440-1E	10.5- 34.1	24.1 – 31.5	0.344 – 0.509	0.446 – 0.600	0.481 – 0.653	-	Manuwa, 2000
TGX 1871- 5E	7.1- 43.7	23.43 – 32.23	0.434 – 0.679	0.4245 – 0.601	0.4243 – 0.6789	-	Manuwa and Afuye, 2004
TGX 1019-2EB	6.7- 47.1	25.87 – 32.45	0.3839 – 0.5774	0.4877 – 0.6249	0.4922 – 0.6876	NAV	Manuwa and Oduganjo, 2005
Unspecified	6.7- 15.3		0.21 - 0.34	0.22 – 0.35*		0.19 – 0.33	Polat et al., 2006
TGX 1448- 2E	9.9 to 39.6	24.2 – 30.2	0.391 – 0.510	0.466 – 0.601		-	Manuwa, 2007
Unspecified	8- 16		0.164 - 0.286				Kibar and Ozturk, 2008
Unspecified	6.92- 21.19	24.56 – 29.93	0.28 – 0.326	0.287 – 0.361		0.262 – 0.307	Tavakoli et al., 2009

MC = moisture content, NAV= not available, PWLG = plywood parallel to grain, PWDG = plywood perpendicular to grain *PLWD = plywood

ESTIMATED VALUES OF SOYBEAN PROPERTIES

Some typical values and models of physical, mechanical and aerodynamic properties of soybean cultivars are reported in this section (Tables 4 to 6). Table 4 shows the effect of moisture content on mass and dimensional properties of some soybean cultivars. Table 5 shows the effect of moisture content on density, sphericity, porosity and terminal velocity of some soybean cultivars. Table 6 shows the effect of moisture content on angle of repose and coefficient of static friction of some soybean cultivars.

GENERAL COMMENTS

It can be seen that the number of soybean cultivars that have been developed around the world is numerous and can be better imagined. However, it appears that very little has been reported in literature concerning physical and engineering properties of such soybean cultivars. Nevertheless, it is obvious that post harvest options or technology are *sine qua non* in order to convert soybean seeds into quality food for human and animal in view of the quality of food nutrition available in the seeds.

REFERENCES

1. ASAE.(2006a) Compression tests of food materials of convex shape. S368.4, 609- 616
2. Braga G.C., Couto S. M., Hara T., J.T.P.A. Neto (1999). Mechanical behaviour of macadamia nut under compression loading. Journal of Agricultural Engineering Research, 72: 239- 245
3. Carman K (1996). Some physical properties of lentil seeds. Jour Agric Engng Res 63: 87-92 Carman K, Ogut H (1991). The determination of porosity rate on different moisture content of several crops. Agric Fac, Univ of Selcuk, Kenya, Turkey 1: 55-62
4. Chung J H, Verma L R (1989). Determination of friction coefficients of beans and peanuts. Trans ASAE 32: 745-750
5. Deshpande S D, N Ali (1988). Effect of harvest moisture on some engineering properties of 'Wheat'. Int AgroPhys 4: 83-91
6. Deshpande S D, Bal S, Ojha T P (1993). Physical properties of soybean seeds. J Agric Eng Res 56: 89-92
7. Dutta S K, Nema V K, Bhardwaj R K (1988). Physical properties of gram. J Agric Eng Res 39: 259-268
8. F A O (1988). Production yearbook. Food and agricultural organization of the united nations, Rome, Italy
9. Fraser B M, Verma S S, Muir W E (1978). Some physical properties of fababeans. J Agric Eng Res 22: 53-57
10. Glass K.M., Delaney D.P., E.V. Santen (2006). Performance of Soybean Varieties in Alabama, USA. Agronomy and Soils Department Series No. 279. Alabama Agricultural Experimental Station
11. Henderson M E (1982). Agricultural processes engineering 3rd edn. AVI Publ Co. Inc, Westport, Connecticut
12. Jha S W (1999). Physical hygroscopic properties of makhana. J Agric Engng Res 72: 145-150 Josh D C, Das S K, Mukherjee R K (1993). Physical properties of pumpkin seed. J Agric Eng Res 54: 219-22.

13. Kibar H. And T. OOzturk (2008). Physical and mechanical properties of soybean. Int. Agrophysics, 22, 239- 244
14. Liu B Y, Nearing M A, Baffaut C, Ascongh J C (1997). The watershed model: III. Comparisons to measured data from small watersheds. Trans ASAE 40: 945-952
15. Mani S., Tabil L. G., and Sokhansanj S. (2004). Mechanical properties of corn stover grind. Transactions of the ASAE, 47, 1983- 1990.
16. Manuwa S I (2000). Properties of soybean 'TGX1440-1E' Annals of Agric Sci. 1 (3) 85-93. Manuwa S I and Afuye G. (2004). Moisture-dependent physical properties of soybean (var TGX1871-5E) Nigerian Journal of Industrial and System Studies , Vol3 (2): 45-53.
17. Manuwa S I and Odubanjo O.O. (2005). Physical properties of soybean (var TGX 1019-2EB) LAUTECH Journal of Engineering and Technology, Vol 3 (2): 88-92.
18. Manuwa S I (2007). Moisture-dependent physical properties of improved soybean 'var TGX1448-2E' J. of Food Sci and Technol, 44 (4): 371- 374.
19. McCabe W L, Smith J C, Harriot P (1986).Unit operation of chemical engineering McGraw Hill Book of Co. New York
20. Mohsenin N N (1970). Physical properties of plant and animal material. Gordon and Breeach, New York
21. Molenda M., Montross M D., Horabik j., and Rose I.J. (2002). Mechanical properties of corn and soybean meal. Transactions of the ASAE, 45, 1929- 1936.
22. Ogut H (1998). Some physical properties of white lupin. J Agric. Eng Res 69: 273-277
23. Ogut H and K Carmam (1991). Determination of coefficient of friction on different surfaces of small grain crops. National Symp Mechanization in Agriculture, Konya, Turkey 471-480
24. Oje K (1994). Moisture dependence of some physical properties of cowpea. Ife J Technol 4 (2): 23-27
25. Polat R., Atay U., and C. Saglam (2006). Some physical and aerodynamic properties of soybean. Journal of Agronomy 5(1): 74= 78
26. Sharma S K, R K Dubey, C K Techchandani (1985). Engineering properties of black gram, soybean and green gram. Proc Indian Society of Agricultural Engineers, 3: 181-185
27. Shephered H, Bhardwaj R K (1986). Moisture-dependent physical

properties of pigeon pea. J Agric. Eng Res 35: 227-234
28. Smith A K, Circle S J (1972). Chemical composition of the seed. *In* soybeans: chemistry and technology, AVI publ Co., Smith A K, Circle S J (eds), Westport, Connecticut,1: 61- 92
29. Sreenarayanan V V, Subramainan V, Visvanthan R (1985). Physical and thermal properties of soybean. Proc Indian Society of Agricultural Engineers 3: 161-167
30. Tavakoli H., Rajabipour A., S.S. Mohtasebi (2009). Moisture-dependent some Engineering properties o Soybean grains. Agricultural Engineering International: the CIGR Ejournal. Manuscript 1110. Vol. XI. February, 2009
31. Thompson R A, Isaacs B W (1967). Porosity determinations of grains and seeds with comparison Pyonometer. Trans ASAE 10: 693-696
32. Uzuner, B A (1996). Soil mechanics. Technique Press, Ankara, Turkey
33. Zou Y. and G H. Brucewitz (2001). Angle of internal friction and cohesion of consolidated gound marigold petals. Transactions of the ASAE, 44, 1255- 1259

INDEX